Thermo-Mechanical Solar Power Plants

Solar Energy R&D in the European Community

Series B:

Thermo-Mechanical Solar Power Plants

Volume 1

Publication arrangements: D. NICOLAY

Solar Energy R&D
in the European Community

Series B Volume 1

Thermo-Mechanical Solar Power Plants

EURELIOS, The 1 MWel Experimental Solar Thermal Electric Power Plant in the European Community

Final report of the Construction of EURELIOS, assembled by the Industrial Consortium for the Implementation of the European Solar Power Plant

edited by

J. GRETZ, A. STRUB and W. PALZ
Commission of the European Communities,
Directorate-General Science, Research and Development

D. Reidel Publishing Company
A MEMBER OF THE KLUWER ACADEMIC PUBLISHERS GROUP
Dordrecht / Boston / Lancaster
for the Commission of the European Communities

Library of Congress Cataloging in Publication Data
Main entry under title:

Eurelios, the 1MWel experimental solar thermal electric power plant of the European Community.

(Solar energy R&D in the European community. Series B, Thermo-mechanical solar power plants: v. 1)
1. Solar power plants-Italy-Sicily. I. Strub, A. S. (Albert S.) II. Gretz, J., 1928- . III. Palz, W. (Wolfgang), 1937- IV. Commission of the European Communities. V. Series.
TK1056.E97 1984 621.31'213 83-26962
ISBN 90-277-1728-1

CIP

Organization of the Contractors meeting by
Commission of the European Communities
Directorate-General Science, Research and Development, Brussels

Publication arrangements by
Commission of the European Communities
Directorate-General Information Market and Innovation, Luxembourg

EUR 9127

Published by D. Reidel Publishing Company
P.O. Box 17, 3300 AA Dordrecht, Holland

Sold and distributed in the U.S.A. and Canada
by Kluwer Academic Publishers,
190 Old Derby Street, Hingham, MA 02043, U.S.A.

In all other countries, sold and distributed
by Kluwer Academic Publishers Group,
P.O. Box 322, 3300 AH Dordrecht, Holland

Printed in The Netherlands

Preface

This book reports on the construction phase of the European solar power plant EURELIOS.

EURELIOS was the first of half a dozen solar power plants in the MW range around the world to enter operation.

It is of the central receiver type employing steam as the heat transfer fluid and has a nominal electric power of 1 MW. The Project was built in Sicily as part of the Solar Energy R + D Programme of the Commission of the European Communities.

Funding was provided by the European Communities, the Governments of France Germany and Italy and the contractors themselves, i.e. ENEL, ANSALDO, CETEL and MBB.

The active involvement of the national utility of Italy, ENEL, the Commission's host partner for EURELIOS, was of paramount importance for the successful completion of the project and should be stressed here.

Many individuals made contributions. The original concept was developed by Prof. FRANCIA who was a leading expert in the field. Dr. STRUB, Director for Non-Nuclear Energy R + D at the Commission of the European Communities (D.G. XII) had the general responsibility for the project, Prof. CHABBAL from France and Dr. KLEIN from Germany very much contributed to its initiation, Dr. CORVI from Italy was actively following the project all the way through its implementation phase.

Implantation was carried out with Dr. GRETZ as the Commission's project leader assisted by General Technology Systems, Ltd from United Kingdom and under the guidance of a very efficient management committee. In this committee the industrial consortium was represented by the following members : Messrs BORGESE, DINELLI (ENEL), TUARDICH, AMADEI (ANSALDO), TREIBER, HOFMANN, SCHOBER (MBB), TOTH, FAURE, DEPAUW, BLANCHARDIE (CETHEL).

It is important to note that EURELIOS is a purely experimental plant. The prime objective of its construction was the technical evaluation of this new and highly innovative conversion process. Its implementation at a relatively large scale - still being much below the full scale for commercial power plants of this type - was necessary because of the complexity of the involved technology.

At this time it has become clear that the construction of EURELIOS was worthwhile. Its value as well as its technical and economical limits could be assessed and much detailed experience has been gained. It is also interesting to note that EURELIOS has become an outstanding example of transnational cooperation among Administrations and among industry in Europe.

W. PALZ,
Head of the EC Solar energy R + D Programme.

Contents

1. Introduction

A "power tower" solar-thermo-electric plant comprises a field of many mirrors controlled as heliostats which focus sunlight onto a receiver mounted on a tower. The resulting thermal energy is transferred by means of a working fluid, or fluids, to an electricity generator - in this project water and a steam turbine driven alternator, respectively. Optionally some means of short term energy storage may be included - in this project high pressure hot water and an additional superheating fluid, molten salt - to allow the plant to continue operation during, for example, the passage of clouds.

The Commission of the European Communities (CEC), DG XII - Research Science and Development, placed a contract for an initial feasibility study (Phase A) between May and October 1976. This contract resulted in the specification of a plant which would provide valuable design and operating experience on a number of novel problems inherent in achieving viable solar thermo-electric plants.

Phases B and C, which ran for three years from November 1977, comprised respectively detailed design and construction of the plant near Adrano in Sicily. Named EURELIOS, it was completed before Christmas 1980, commissioning operations having already started.

In January 1981, during the formal acceptance process of the Commission and ENEL (Ente Nazionale per l'Energia Elettrica, the operator of the plant) the project moved into its fourth phase, D, during which the output power is to be increased to the design figure and an experimental programme started.

The work was sponsored jointly by the CEC and the governments of Italy, France and F.R.Germany. The Commission's contract for design and construction provided for the payment by the Commission of half the expenditures incurred.

The organisation of the project is shown in Fig. 1.1. The plant has been built by an industrial consortium consisting of ENEL and ANSALDO-AMN (now Ansaldo Impianti) - Italy, CETHEL - France, and MBB (Messerschmitt Bölkow-Blohm GmbH) - F.R.Germany. ENEL also acts as host-partner, providing free of charge the site and supporting site services such as security and having been responsible for ensuring the safety and legality of the complete plant. Within the Consortium, ENEL was responsible for civil works interfacing with the Sicilian electrical grid, the design of the control system and the definition of the operation procedures of the plant. ANSALDO supplied the tower and receiver, the steam cycle, hot water/steam storage and associated electrical and pneumatic controls. CETHEL's responsibility was to provide approximately half of the heliostat field and associated monitoring and controls, the molten salt thermal storage system and the internal electrical power system (alternator, main transformer and auxiliary power distribution). The other half of the heliostat field (rather more than half in terms of numbers of heliostats but, since its heliostats are smaller than those of CETHEL, less than half the total reflecting area) was supplied by MBB. Each company provides special electrical conversion and, where necessary, backup equipment, drawing power from Eurelios itself or from the grid. General Technology Systems Ltd., has been retained separately by the Commission, through its Belgian subsidiary, GTS SA, for assisting the Commission's project leader in the monitoring of the work and to act as the Commission's agent during acceptance testing at the end of Phase C.

A Helioelectric Plant Consultative Committee (HPCC), which consists of members proposed by the Commission (3) and the participating member states (3) and which has observers from the Consortium and from ENEL (as host), advises the Commission on overall policy and strategy of the project, on matters relating to the project needing a decision by the Council of the European Communities and on the passage from Phase B to Phase C and commented on the contractual reports of the Consortium.

This report first outlines the original requirements of the plant, and briefly describes it (Chapter 2) and then describes the plant in detail, subsystem by subsystem (Chapter 3).
Chapter 4 outlines the operating procedures and the report ends with a look forward into Phase D.

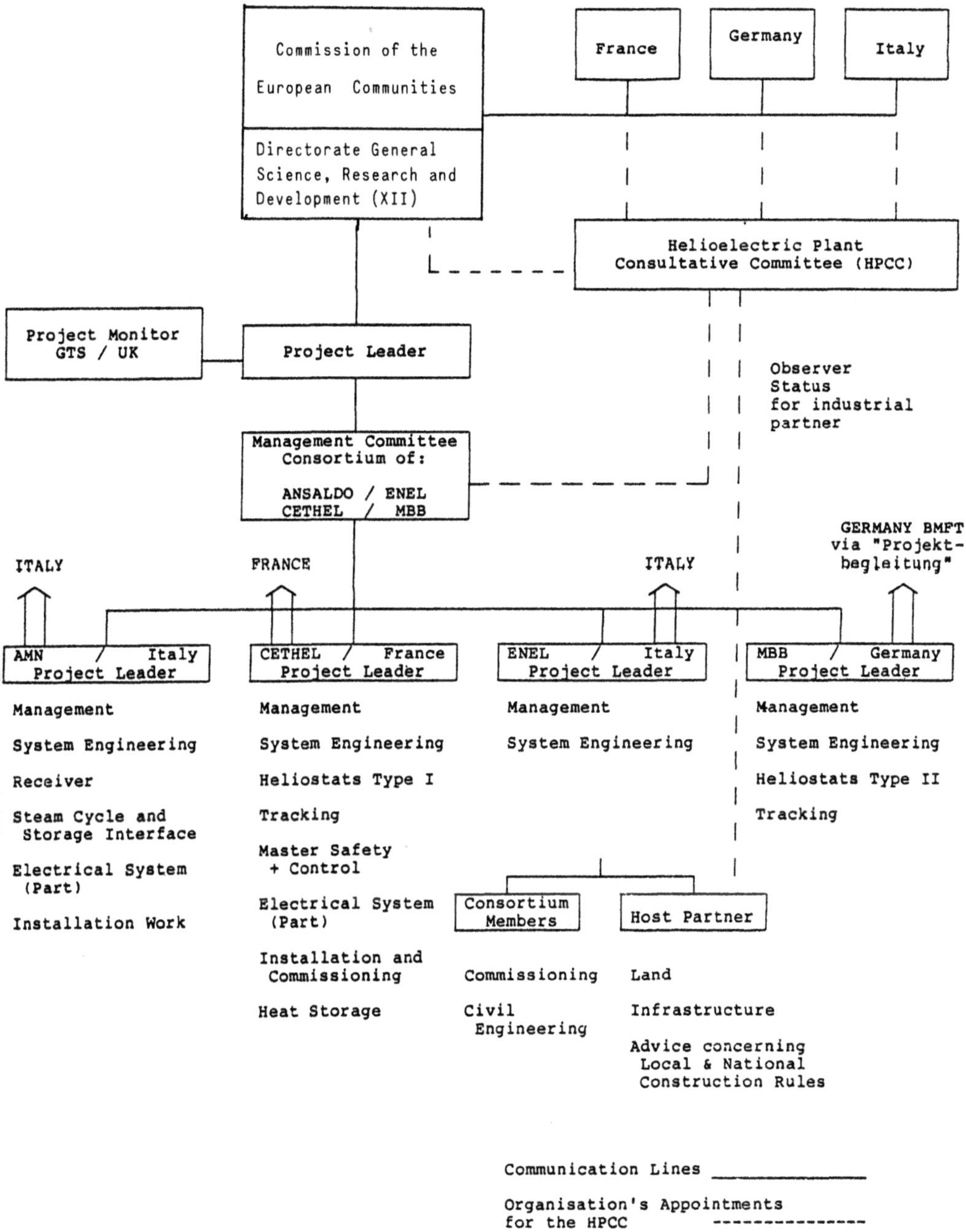

Figure 1-1 - EURELIOS Project Organisation (Phase B and C)

2. System Design Considerations

2.1 BASIC REQUIREMENTS

Ideally, all parameters of the plant could be considered to be adjustable to maximise cost-effectiveness according to system level requirements: e.g. the increase of temperature of the receiver could lead to a higher efficiency of the power conversion cycle and to a reduction of number of heliostats but requires an increase in pointing accuracy. In practice, however, the opportunities for parameter variation are limited by available or preferred technology.

For the EURELIOS plant, once the site had been selected, the following features were specified (although many were slightly altered during the design phase as a result of optimisation).

. Nominal power output: 1 MWel at an insolation of 1 kW/m^2 at equinox noon at the site of Adrano/Sicily with the following geographical coordinates:

  37.6466° north

  14.8034° east.

  The site slopes down about 4° from north to south and 2.5° from east to west. The elevation above sea level of the foot of the tower is 214 m. The heliostat field to be north of the receiver.

. Tolerable wind forces:
  - nominal output power to be available up to 18 km/h, the maximum tower deflection being ± 10 cm;
  - operation (reduced power) to be possible up to 50 km/h;
  - survival (no permanent deformation) up to 120 km/h (130 km/h for the tower, a figure which was, in the event, adopted for most of the plant).

. Utilisation of a steam cycle and cavity type boiler with anti-radiating structure (made of Pyrex glass, design of Prof. Francia) using water/steam directly as heat transfer and working medium with a flow rate of up to 5300 kg/h allowing growth over the then nominal conditions of 4860 kg/h. Nominal steam conditions are 510°C, 64 atm. Receiver aperture about 4.5 m diameter (amended as a result of early optimisation work from an initial 4.2 m x 4.2 m square aperture in a 5 m cube). Incoming radiation flux into the aperture is 4800 kW at the nominal design point.

. A steel tower to support the receiver, about 50 m high (centre of aperture) but this increased to 55 m, again resulting from optimisation during design.

. CETHEL heliostats each with 52 m^2 reflective surface and MBB heliostats with 23 m^2, each giving a reflected beam accurate to ± 4 mrad

(1 sigma). The CETHEL field section to give at least 57% of the total requirements; the MBB section at least 43%.

. Thermal storage to allow the plant to continue operating for 30 minutes without insolation.
. A conventional electrical system with a 1.1 MW alternator (minimum), connected to the national grid's Contrasto hydroelectric power station via a 1.0 MW output transformer (minimum). To have automatic emergency shut-down and appropriate emergency devices to ensure the safety of the plant despite equipment failures or loss of power.
. Centralised control.
. The whole plant to conform to the relevant Italian law and regulations.

Priority has been given to the experimental character of the plant. Hence although automatic start-up and shut-down are ultimate and important objectives, operation will involve considerable manual intervention. Similar comments apply to recovery from abnormal, including emergency, conditions subject, of course, to the need for automatic initiation of activities needed to ensure the safety of personnel and plant.

2.2 SYSTEM DESCRIPTION OF THE PLANT

Figure 2.1 is a simple schematic of the EURELIOS plant, the principal characteristics of which are summarised in Table 2.1.

The individual subsystems are described in detail in Chapter 3 of this volume. This section explains the overall configuration so that the individual sections of Chapter 3 can be read in isolation by those interested in only limited parts of the plant.

2.2.1 Heliostats

Sunlight is intercepted by the heliostat "field" and reflected into the receiver (boiler) just to its south. The field is split into two roughly equal "subfields", East and West, supplied by MBB and CETHEL, respectively. The subfields are each divided, for control purposes, into four independent "groups". Each heliostat has its own microprocessor-based local electronics unit which is controlled by, and reports its status to a "Central Unit" (one each for CETHEL and MBB). The Central Units can be used manually to control individual heliostats, groups or the corresponding subfield. The Central Units are also controllable by, and report to, the plant's Generator Control System (GCS) for routine and emergency control at group level and above only.

The CETHEL heliostats also contain emergency batteries with several times the amount of energy needed to move the heliostats to a safe position in the event of loss of the electrical supply.

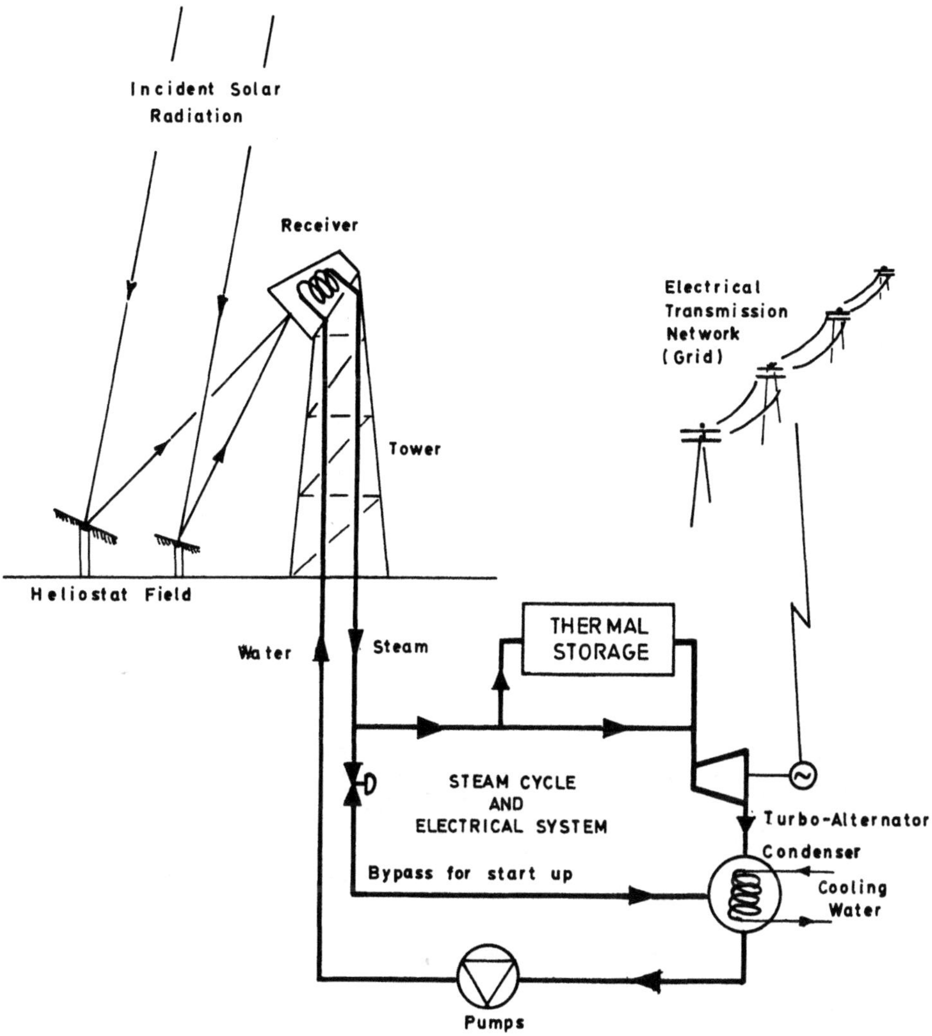

Figure 2-1 – EURELIOS Concept

1 MW_{el} SOLAR POWER PLANT OF THE EEC

Technical Parameters

General characteristics

Experimental plant of the central receiver multiheliostat type
Location : 37.6° North; 14.8° East (Adrano, Sicily, Italy)
Design point : equinox noon and assumed insolation of 1,000 W/m^2
Power generated : 1 MW_{el} into existing grid at design point

Design parameters (power figures at design point)

Heliostat field	: 4800 kW_{th} to receiver - 6202m^2
Heliostats	: two types, two axis controlled, focusing, overall inaccuracy ± 4 mrad (1 sigma)
Cethel type	: ca. 52 m^2, 8 modules, 70 heliostats
MBB type	: ca. 23 m^2, 16 square elements, 112 heliostats
Receiver/tower	: Cavity type receiver 4.5 m aperture at 55 m height, 110° inclination
Receiver outlet steam conditions	: 512°C, 64 atm, 4860 kg/h
Steam cycle	: Turbine connected to receiver (no intermediate heat exchanger)
Nominal power	: 1100 kW_{Mech} with steam of 510°C, 60 atm
Feed water temperature at receiver inlet	: 36°C
Cooling water temperature	: 25°C maximum
Thermal storage	: Reduced electrical output for ca. 30 min.
Capacities	: Vapour 300 kWh; Hitec : 60 kWh
Equipment	: Pressurised water reservoir for 4300 kg vapour (reducing from 19 to 7 bar) Two storage tanks, containing about 1250 kg Hitec salt (overall capacity) Heat exchangers for 64 bar, 480°C and 19 bar, 410°C steam
Electrical system	: Alternator 1500 KVA giving internal power and 1000 kW for external users

Transformers
Emergency power supply
Interface to grid : equipment to connect transformers to public grid
Steam cycle control equipment
Command, operation and monitoring is centralised in the control centre

Table 2.1 - Technical Parameters, Summary

2.2.2 Receiver

The ANSALDO receiver is a stainless steel, dualchannel, once-through, cavity-type steam boiler located at the top of a steel tower. Output steam conditions are controlled by adjusting the feedwater flow rate and by the use of attemperator sprays within the boiler. It is equipped with pressure, temperature and thermal flux sensors which allow the control system to make the appropriate adjustments.

The design concept, due to the late Prof. Francia, includes the use of a Pyrex glass anti-radiating structure inside the receiver to reduce losses due to re-radiation and convection. The Pyrex is to be installed after experience without it has been obtained, in order to verify the effect of the anti-radiating structure.

The receiver is protected in various ways against damage due to possible electrical or pump failure or misalignment of heliostats (by ceramic bricks on the rim, by a protective casing, by an over-pressure release valve, by an emergency feedwater supply system, etc.).

2.2.3 Steam cycle

The steam cycle built by ANSALDO is made of stainless steel. Demineralised water is pumped at pressure into the receiver where it is boiled and superheated before being passed for use in one or more of three ways:
i) to drive the turbine for electrical power generation;
ii) to heat the salt of the thermal storage system;
iii) to heat the water for the thermal storage system.

During start-up, when the system is being pressurised and heated, or in emergency, steam is sent through a turbine bypass valve and desuperheater before being passed via a flash (sparge) tank to the condenser for recirculation. In normal circumstances the electrical generation load on the turbine is automatically adjusted to make full use of the available steam despite variable insolation, the control system of the steam cycle adjusting feedwater flow rate to maintain steam pressure and temperature.

The steam cycle control system, built by ANSALDO according to a functional design of ENEL, has as one vital role the protection of the receiver against damage by over-heating. In addition to having duplicated electrical supplies and duplicated pumps the steam cycle includes an emergency pressurised feedwater storage system which operates in the complete absence of the normal feedwater supply until the receiver is free of insolation.

2.2.4 Thermal storage system

This system stores heat in pressurised, saturated steam/water and molten salt, its purpose being to keep the plant operating for half an hour without insolation during cloud passing. The salt, contained in a system provided by CETHEL, is charged (heated) by pumping it from a "cold" to a "hot" tank through a heat exchanger taking heat from the main steam supply. The heavily insulated steam/water accumulator, part of ANSALDO's equipment, is charged by first partially filling it with cold feedwater and then bubbling (condensing) main steam into it until the desired working pressure is reached. During discharge the hot salt is pumped through a second heat exchanger, giving energy to superheat the saturated steam from the accumulator, this steam then being fed to the turbine.

2.2.5 Electrical systems (CETHEL and ANSALDO)

During normal operation, the plant generates electrical power in an alternator driven by the steam turbine. Part of this power is used (via a transformer) at 380 V to run the plant, the greater part being sent (via a second step-up transformer) to the nearby Contrasto hydroelectric power station where it contributes to the Sicilian grid. During start-up or maintenance, power comes normally by the same route from Contrasto. A second, independent, line brings in power to operate the auxiliaries if the normal supply is for any reason not available. Services essential for the emergency survival of the plant can also obtain electrical power from a bank of lead-acid batteries.

Each subsystem itself carries out any necessary further power conditioning on the 380 V three-phase supply distributed around the plant.

2.2.6 Control system

The plant is centrally controlled from a room overlooking the machine room. The control room contains the main control console for operating the plant, vertical boards, in front of the operator, displaying a synoptic of the process, the alarms, meteorological data, plant thermodynamic and electrical data.

Near the console are the desks and computers for detailed monitoring and control of the two heliostat subfields. Beneath the control room is located all the electrical switchgear whereas the control apparatus are installed in an adjacent equipment room.

2.2.7 Central services

The facilities for operating the plant are mostly housed in a building to the south west of the tower and receiver. There are in addition ancillary buildings housing spare parts, workshop facilities and exhibition, conference and office accommodation for visiting specialists not involved in the day-to-day operation of the plant. Meteorological monitoring is carried out at a number of locations around the site.

2.3 THE NOMINAL POWER BUDGET

		kW
Insolation intercepted by heliostat field, gross (1)		6200 (2)
Less shading, cosine, misalignment, reflectivity and blocking losses (3)		1400 (4)
Power into receiver (visible and thermal radiation)		4800
Less reflection, reradiation and convection losses (inc. pipework)		416 (5)
Net power absorbed by working fluid		4327 (6)
Less losses in steam cycle		3227 (6)
Shaft power from turbine		1100 (6)
Less alternator losses		50 (7)
Electrical output from alternator		1050
Less consumption and losses in auxiliaries (normal operation):		
steam cycle	32 kW	
CETHEL field	4 kW	
MBB field	4 kW	
central controls	5 kW	
other	5 kW	50
NET OUTPUT OF THE PLANT		1000

Notes:

1) Nominal, 1 kW/m^2 insolation but with mirrors normal to sun line, neglecting space between mirrors. All figures rounded.
2) Based on total mirror area: 70 x 51.8(CETHEL)+ 112 x 23 (MBB) = 6202 m^2.
3) Shading and blocking are zero at equinox noon.
4) Figure obtained by differencing from the one below.
5) Figure obtained from the one below, assuming 91.3% efficiency.
6) Figure obtained from AMN "nominal output" data for turbine. Net power absorbed is obtained by differencing enthalpies of steam-to-turbine and feedwater for the nominal flow rate.
7) Based on 95% efficiency of alternator for 1200 kW load.

2.4 SAFETY

Safety is particularly important in an experimental facility to which the public has access but which involves high voltages, high pressure steam and highly concentrated sunlight. The primary safety objective must be to avoid personal injury. Following this closely are damage outside the plant and damage to the plant itself.

The main unconventional risks to both personnel and equipment result from concentrated sunlight. It was established early in the project that light reflected from the receiver would not be a hazard but that care would be necessary to avoid risk from the focussed light from the heliostat mirrors. Light from a single heliostat is only a hazard, causing serious dazzling and possibly eye damage, to people near the focus of the heliostat. Since no heliostat has a focal distance of more than a nominal 200 m, such risk is limited to the immediate vicinity of the plant itself. To extend this limitation to the whole heliostat field rather than its applying simply to one heliostat, the concept of a Safety Trajectory was introduced. A vertical line is defined, adjacent and just to the east of the tower. The beam centre of every heliostat is constrained to follow this line wherever it is in transit between the ground and a Standby Position immediately east of the centre of the receiver aperture. Since heliostats are by this means focussed within the plant, no pair is ever likely to converge on any distant object and in addition the individual heliostat beams would even then be well out of focus themselves.

However, there still remains at risk the immediate vicinity of the plant, either from a single heliostat or from numbers of them. An analysis has been carried out to find the effect of any failure of a heliostat which causes it to stop or move out-of-control, so as to illuminate the surrounding area. The value of 3 kW/m^2 was selected as one which was very dangerous for the human eye.

To compute the iso-flux the following hypotheses were made:

a) The normal direct radiation intensity can be written as follows:

$$I = \exp(\sin E/\sin E_{1200} - 1)$$

where E is the angle between the solar radiation and the horizontal plane at the time and the day considered, E_{1200} is the value of the above angle at noon on the day in question.

b) The linear dimension of the heliostat's focus, due to astigmatism, increases as the inverse of the third power of the cosine angle between the incoming radiation and the normal to the heliostat ($\cos \phi/2$).

c) The single mirrors of the heliostat have the same focal length as the heliostat itself.

d) The reflected radiation is uniformly distributed, i.e. any orthogonal section of reflected beam has a constant power density.

e) The iso-flux lines are on a horizontal plane including the top of heliostat's frames.

The boundary obtained through the above conditions is approximate but is sufficient to give a general indication of the area within which care would have to be taken in the event to such a heliostat failure. Even in the case shown in the figure, a particularly unfavourable one, only a very limited area beyond the site boundary is affected.

Within the site, particularly at the foot of the Safety Trajectory, there would be risk whenever there were heliostats in transit and so a protective wall (shown in the figure) has been built to protect personnel near the foot, and to the east, of the tower. Further, personnel are not normally allowed on the tower or in the field when heliostats are in use.

Because of the Safety Trajectory concept the remaining danger is limited to the effects of the focussed spot near the receiver - either from poor focussing or misplacement of the spot as a whole. In the former case, which may occur whenever the sun's position is distant from that at which a heliostat was focussed, the beams spread, illuminating the sides of the receiver rim and possibly the structure of the tower. Structural illumination is minimal but an unprotected rim may be overheated. However, since the whole spot is formed from the full heliostat field, the individual members of which are individually controlled, failure only affects the whole field in the event of power failure, in which case the CETHEL heliostats still move out of the receiver under battery power. The MBB heliostats remain stationary. Provision has to be made to protect not only the receiver pipework (Eurelios includes an emergency feedwater system not requiring electrical power) but also to protect the eastern rim of the receiver while the spot moves eastward due to the motion of the sun. In fact the entire rim of the receiver is protected with ceramic bricks.

People are protected against the conventional risks in conventional ways: by excluding them from dangerous areas by using guards, fences, passes, closed cabinets, colour codings, warning notices, use of protective head gear. The plant is of course built to satisfy Italian national safety regulations and includes duplicate and safety-specific equipment in many places, to protect the operation of the plant itself from disruption.

3. Subsystem Design Information

3.1 HELIOSTATS

The heliostat field is required to provide a maximum of power into the receiver aperture. Fig. 3.1.1 illustrates the most important effects influencing the amount of radiation flux finally entering the receiver aperture. With the help of computer programs numerical calculations of the radiation power were performed for various arrangements of heliostats, leading finally to the field layout shown in Fig. 3.1.2.

Apart from the primary requirement, i.e. optimum power input into the receiver, there are several secondary requirements, which constrain the layout of the EURELIOS heliostat field.

For both types of heliostats, i.e. CETHEL and MBB, it was required to study the time variation of the image of the sun (alignment, quality of mirrors, etc.) and its effects on optical efficiency, both for heliostats close to the tower and more distant from the receiver. It was therefore decided to arrange the heliostats in such a way that there is a west-field (CETHEL) and an east-field (MBB). This gives similar elevation (and azimuth) angles in the west-field in the morning, as in the east-field in the afternoon. As CETHEL is to supply 57% of the nominal receiver input power, some of the CETHEL heliostats are placed east of the exact north-south line.

A dense positioning of heliostats is preferred, to minimise use of land and to gain as much radiation as possible: unavoidable imperfections in reflection lead to increased spillage of radiation for more distant heliostats. Minimum spacings of 11.7 m for the CETHEL heliostats and 7.8 m for the MBB's ensure that adjacent heliostats never touch.

To simplify the civil work, it was decided to arrange the necessary access roads within the heliostat field in the east-west direction and to have one additional road north-south, approximately at the separation line between the two types of heliostats. Accordingly, the heliostats have been arranged in east-west rows as shown in Fig. 3.1.2, between 25 m and 197.7 m north of the receiver. In case of future enlargement, additional heliostats might be put along the same east-west rows, just increasing the number of units per row.

Within the rows, i.e. in east-west direction, the minimum distance between heliostats is adopted. Heliostat rows more distant from the tower are further apart so as to minimise blocking of reflected light by heliostats

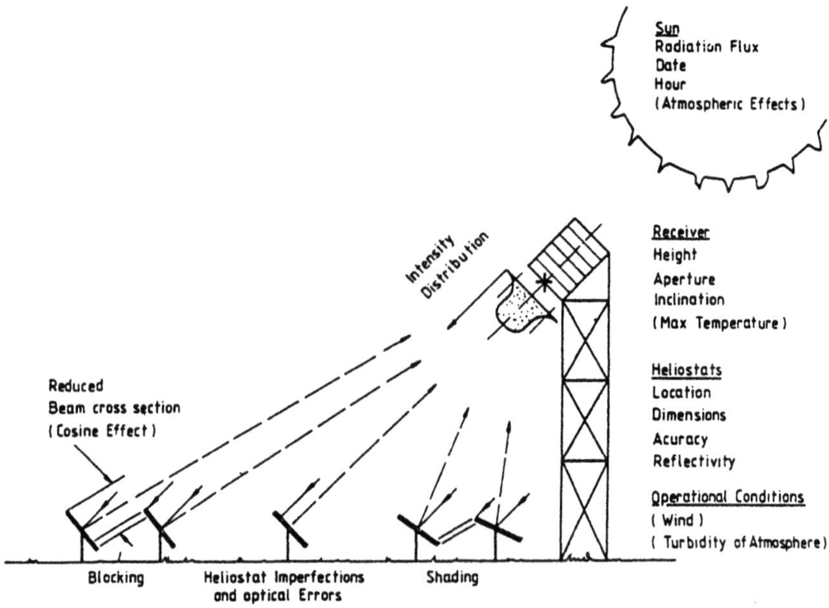

Figure 3.1.1 - Radiation Power Collected in a Central Receiver

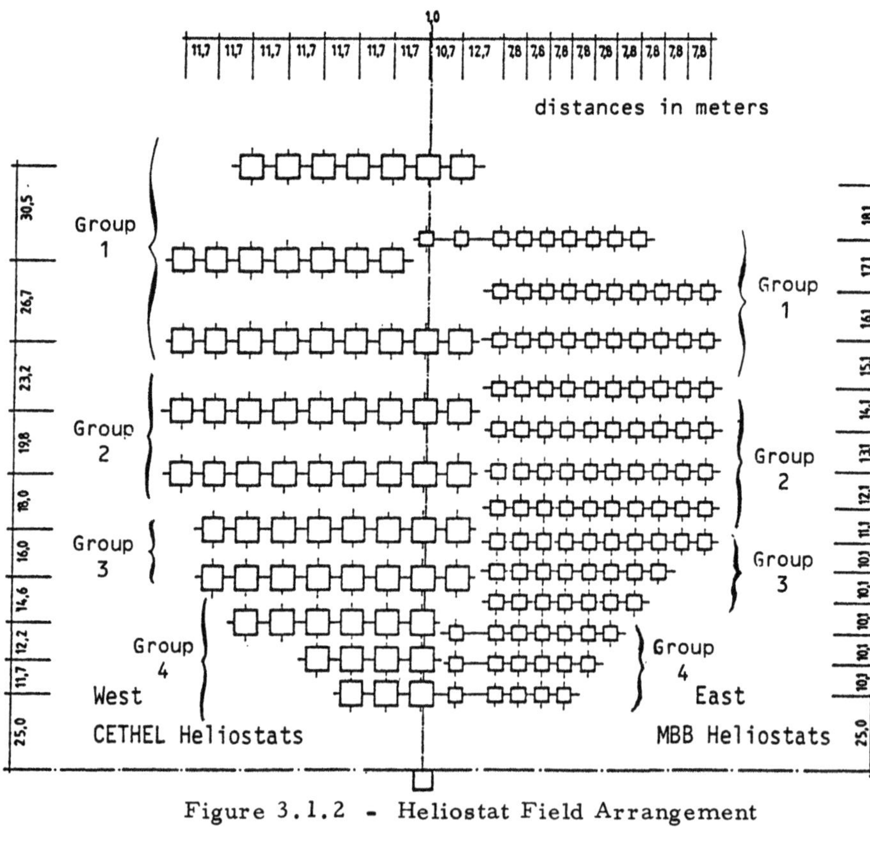

Figure 3.1.2 - Heliostat Field Arrangement

closer to the tower. Whereas the first rows of heliostats in the CETHEL section are as close as the above minimum distance (i.e. 11.7 m) the distance between the first rows of MBB heliostats is increased to 10.1 m (instead of 7.8 m) to allow easy passage for vehicles. Energy calculations have shown that this has an effect on the power output of the MBB field well below 1%.

The final arrangement has been evaluated by a mathematical model of the field with which the energy flux into the receiver was evaluated, taking into account the different geometrical effects indicated in Fig. 3.1.2 as well as the specified gaussian distribution of the direction of the reflected beam about the ideal direction of ± 4 mrad (1 sigma).

The result was that with the 70 CETHEL heliostats and the 112 MBB heliostats arranged as shown, the required radiation power of 4800 kW would be provided to the receiver aperture at nominal insolation conditions, viz. equinox noon, solar time, assuming a solar incident flux of 1 kW/m^2. This arrangement is optimum in the sense that slight variations of the distance of the heliostat rows do not change the power output dramatically but show the tendency that both larger row distances and shorter row distances lead to slightly reduced power outputs. There is neither blockage nor shadowing at equinox noon.

Averaged over the whole MBB field at equinox noon, the reflectivity is 85% and cosine effects and spillage due to imperfect reflection cause a loss of 5% (i.e. 95% "field performance") leading to an overall efficiency of 81%. With a total of 2576 m^2 of MBB reflective surface, this leads to a nominal contribution of 2086 kW, which corresponds to 43% of the required 4800 kW. Similarly for CETHEL with an 80% reflectivity, 3626 m^2 of reflective surface and 95% field performance, the contribution is 2756 kW, i.e. the balance of 57% required, with 1% margin. For single heliostats, the values differ according to their positions and they depend on time.

No system was installed during phase C that could allow any quantitative measure of the actual performance of the heliostats.

Only a 5 m square auxiliary target marked with a half-meter grid was located immediately below the receiver. This allows focussing and alignment of individual heliostats to proceed even when the receiver is being used for routine generation of power.

The east-west split of the field is largely irrelevant to the receiver and steam cycle. However, it is operationally desirable, during start-up and in emergencies, for the field to be subdivided. So, the MBB and CETHEL fields are each divided into four groups which are normally commanded as pairs (CETHEL group 1 with MBB group 1) by the General Control System (GCS).

Table 3.1.1 gives the basic requirements which had to be met by the heliostats. The following sections describe the approaches adopted by the two companies, Table 3.1.2 and Figs. 3.1.3 and 3.1.4 giving general information in comparative form.

Wind velocity	Full Power	18 km/h (5 m/s)
	Operational limit	50 km/h (14 m/s)
	Survival	130 km/h (36 m/s)
Time to reach stowage position (time allowed for 180° movement in elevation)		max. 30 min.
Environment	Temperature	0 to +40°C
	Qualification limits	-10° to +50°C
		Sandstorm Thunderstorm
Reflected beam accuracy at 18 km/h wind		± 4 mrad (1 sigma) Reflected beam

Table 3.1.1 - Heliostat Design Requirements

The most noticeable differences between the two designs of heliostat are their sizes and their back structures. MBB justified its layout as follows:

- development should be possible without too high a risk and within the given time;
- the heliostat should be optimum for relatively small, syy 1 MWel, central receiver power plants and not primarily designed for larger plants;
- the heliostat should be simple in terms of mounting (weight and dimensions) and transport. This might be particularly important for sites in developing countries;
- components should be available off the shelf. This was particularly critical with respect to gears: larger gears would have had longer supply time including additional development time;
- aberration losses can be kept smaller with smaller heliostats.

CETHEL's design was conceived in the context of the established French national programme, involving somewhat larger stations than Eurelios.

Both manufacturers adopted horizontal (elevation) and vertical (azimuth) tracking axes and a nominally horizontal, face down, stowage position for their heliostats. The heliostats can be driven at two speeds: fast to reach a desired position; slow during tracking of the sun. Sun tracking is open loop to the extent that it relies on computation of sun (and hence heliostat) directions as a function of time without direct measurement of the sun or of the reflected beams. There are, however, differences in the implementation of tracking and, from the safety point of view, they differ in that only CETHEL's heliostats have battery-powered emergency supplies available to drive them in the event of a power failure.

	CETHEL	MBB
REFLECTING PANELS		
Configuration (each heliostat)	8 modules of 6 mirrors of 1.8m x 0.6m = 51.8m^2	16 mirrors of 1.2 x 1.2m = 23m^2
Weight	2700 kg.	800 kg. (27 kg/mirror element)
Mirror elements	6mm backsilvered float glass	3mm backsilvered float glass in glass/glass or glass/metal foam-filled sandwich
Nominal reflectivity (clean)	0.8	0.85
Overall panel width	8.84m	5.63m
Overall panel height (top to bottom)	7.34m	5.01m
Height of elevation axis	4.20m	2.97m
Ground clearance	0.52m	0.47m
DRIVE UNITS		
Panel rotation speeds (azimuth)	12°/min fast; 0.6°/min slow	4.79°/min fast; 0.48°/min slow
Panel rotation speeds (elevation)	" " "	6.28°/min fast; 0.63°/min slow
Range of rotation	±120°Az; +75° -90° El.	± 120°Az; ± 90°El
Weight	1100 kg.	600 kg.
COLUMN		
Construction	Concrete Cylinder, integral with foundation	Steel tube, bolted to foundation
Dimensions	Ø 0.850 x 3.92m	Ø 0.406 x 2.62 m
Weight		180 kg.
FOUNDATION		
Construction	Circular concrete cylinder	Double, circular concrete cylinder
Dimensions	2.750 x 0.6m	0.800 x 0.32m on 2.50 x 0.5m base
NOS. OF H'STATS	70	112
TOTAL MIRROR AREA	3626 m^2	2576 m^2

Table 3.1.2 - Numerical Comparison of CETHEL and MBB Heliostats

Figure 3.1.3 - Heliostats Tracking into Receiver CETHEL (Upper)
MBB (Lower)

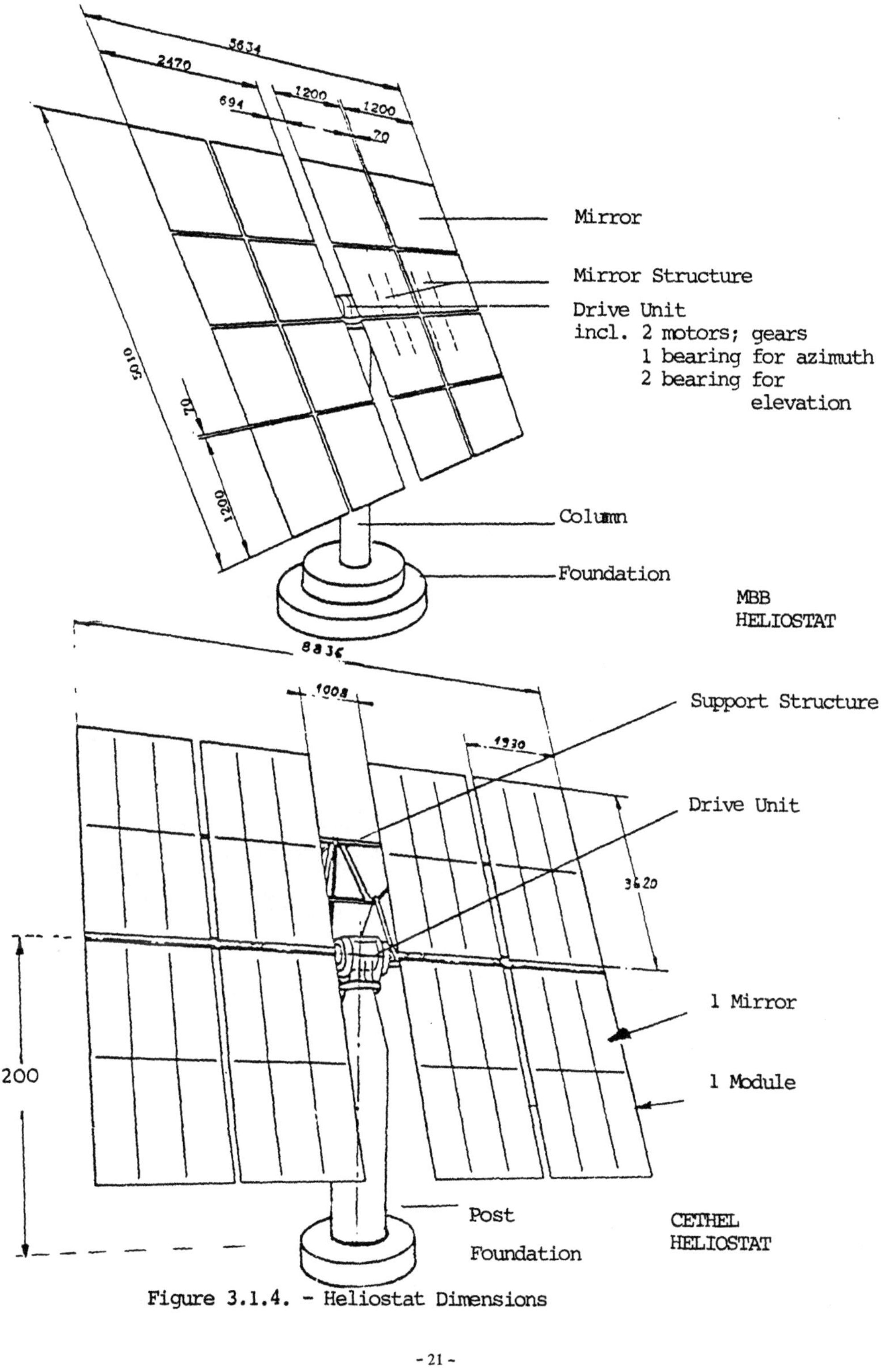

Figure 3.1.4. - Heliostat Dimensions

Heliostat Number	Focal length of module m	Focal length of panel m
1 - 2	80	57,5
3 - 4 - 5		63
6 - 8		69,1
7 - 9 - 10		73,6
11 - 14 - 15 - 16		80,8
12 - 17	100	84,5
18		87,8
13 - 19 - 22 - 23 - 24		93,5
20 - 25 - 26		100
21 - 27		107,5
30 - 31 - 32	140	108,9
28 - 33		110,9
29 - 34 - 35		118,5
36 - 39 - 40 - 41		126,5
37 - 42 - 43		131,1
38 - 44 - 45		139,6
46		148
47		151,9
48 - 49 - 50	200	148
51 - 52		151,9
53 - 54		159,2
55 - 56		169,5
57 - 58 - 59		176,6
60 - 61		182,9
62 - 63		191,9
64 - 65 - 66 - 67		203,6
68 - 69 - 70		210,8

The exact distance between heliostat and receiver aperture is never different from the adjusted focal length of the panel more than 3%.

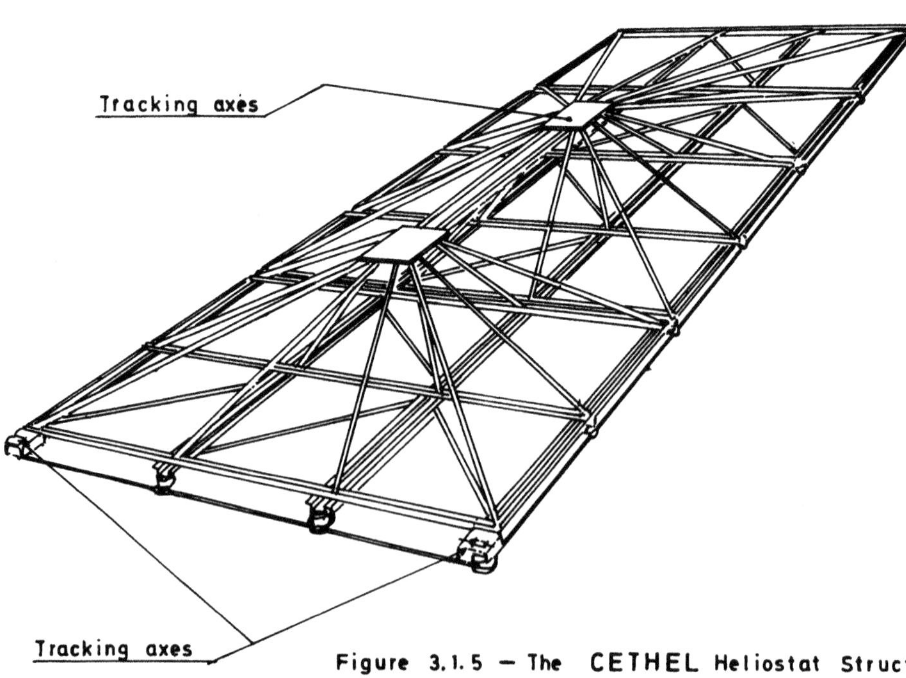

Figure 3.1.5 — The CETHEL Heliostat Structure

Figure 3.1.6 - CETHEL Heliostat Assembly

MBB Installation
Heliostat

MBB Focus on
Auxiliary Target

MBB
Mirror Element

Figure 3.1.7 - MBB Heliostats

3.1.1 Reflecting panels

CETHEL reflecting panels

The CETHEL mirrors are supported by an iron braced-framework structure (see Figs. 3.1.3 and 3.1.6) composed of:

. 8 modules, 3.62 x 1.93 m galvanised iron structure, each of them supporting 6 elementary mirrors 1.8 x 0.6 m;
. a rear structure, made of square section iron tubes, supporting the modules and fitted to the horizontal (elevation) axis of the drive unit.

The mirrors are clamped to the modules via rubber fittings whose elasticity allows for construction tolerances and expansion.

The CETHEL panel shape approximates a spherical mirror with a focal length equal to the distance between the panel centre and the receiver aperture.

First, the modules are shaped so that the 6 mirrors strips are bent cylindrically and canted to be tangent to the ideal sphere. Four types of modules are used in the whole field (sphere radius of 160, 200, 280 and 400 m corresponding to focal lengths of 80, 100, 140 and 200 m). Second the modules are oriented on the rear structure by 3 adjustable supporting bolts so that they focus at the receiver aperture. Again, to avoid too many adjustment parameters, the focussing is designed to be the same for a number of heliostats within a certain range (3%) of actual distance to the receiver (Table 3.1.3).

The CETHEL modules were completely manufactured, welded and galvanised in the workshop in France and transported to the site by train (560 modules, 24 in each wagon). The rear structures were prefabricated, the tri-dimensional structure being divided into flat elements, to be welded and painted outside. The shaping of the modules was also done at the site; aluminium blocks are first riveted to the structure and then cut in height with special tooling in order to give the spherical support to the mirror glasses.

The final erection work consisted of:

. fixing mirrors on modules;
. fixing rear structure to the drive unit, at ground level;
. fixing modules on rear structures, face down, horizontal;
. adjusting the focussing of the panel. The appropriate orientation is given to each module, using the horizontal reference, with water levels and wedges;
. transporting the complete heliostat and putting it on its post in the field with a mobile crane.

MBB reflecting panels

The basic mechanical configuration of the MBB heliostat is shown in Fig. 3.1.7 illustrating the structure before the mirrors are fixed to it. The supporting frames for the mirrors are attached to the drive unit by bolted flanges. All structural parts are finished with weather resisting paint.

The mirror elements are spherically shaped so that they form a sun image at a nominal 200 m distance, about the largest distance between heliostats and the receiver aperture. This use of one focal length may be contrasted with the four different values adopted for the larger CETHEL units.

Most of the MBB mirrors have a core of PVC-foam and an 0.6 mm galvanized steel backing. During the project a second type of mirror was developed, the most visible characteristics of which are that both back and front consist of glass and that the core material is not homogeneous but consists of two components. About 60 of those mirror elements have been mounted. This mirror design is fully compatible with the requirements for EURELIOS.

The mirrors are fixed to the support structure by means of 3 screws. This allows alignment of each mirror by canting it with respect to the structure and to its neighbour mirrors, using the sun. To find the optical axis of the heliostat, an auxiliary mirror is mounted on a bridge type fixture between the two supporting structures and aligned parallel to the elevation axis and to the support structure. Then the heliostat is oriented such that the auxiliary mirror generates a sun image at the auxiliary target of the tower. The 16 mirror elements can then be aligned such that their sun images coincide with that of the auxiliary mirror. After this activity the final readings of the encoder outputs are recorded. Fig. 3.1.7 also shows a focus of a complete heliostat at the auxiliary target.

MBB's calculated optical errors in an 18 km/hr wind are (1 sigma values):

mirror contour	1.47 m rad
mirror alignment	0.44 m rad
temperature effects	0.41 m rad
structural deformations	0.23 m rad
Total optical errors (1 sigma)	1.60 m rad

The total sun image angle including 95% of all reflected energy is 13 m rad.

3.1.2 Drive units, columns and foundations

CETHEL drive units, columns and foundations

The CETHEL mechanical unit, which supports and rotates the reflecting panel in azimuth and elevation, is located at the centre of the panel and fixed on top of the reinforced concrete post (Fig. 3.1.6). It includes all the necessary bearings, gears, motors, encoders, switches, etc.

It is a compact mechanism and so can be tightly sealed for transport, is very easy to link with the reflector and with the post, and can be completely sealed to protect the internal elements of the mechanism from dust, sand and humidity. Since the lower part of the reflector is well clear of the ground, the presence of vegetation, stones, snow, etc., is not important.

CETHEL's elevation and azimuth motors are identical. Their main characteristics are:

. D. C. motors with permanent magnets;
. 2 directions of rotation;
. operating power: 90 W at fast speed; 40 W at slow speed, in tracking;
. rated speed: 1500 rpm ($\pm$ 5%);
. constant torque of 0.6 Nm $\pm$ 10^{-3} Nm from 80 rpm to 1500 rpm;
. input voltage: 240 V at 1500 rpm;
. rated current: 6 amp. (about);
. efficiency: 60%;
. closed motors, drop proof;
. protection: IP 44.

The 1:11,200 reduction gears, also identical for azimuth and elevation, consist of parallel trains of gear-wheels driven by a worm-gear, all carefully manufactured with special steel to minimise backlash or deformations. The total error, including the effects of wind and weight, is limited to 1 mrad (3.4 arc min).

The azimuth axis relies on a toothed orientation ring bolted on one side to the pillar and on the other to the cast support of the mechanism. The ring is driven by the slow speed pinion of the (azimuth) reduction gear, the gear ratio being 32/129.

The elevation axis is implemented as a tube connected to the two wings of the reflector by means of two flanges. It also uses a toothed orientation ring driven by the slow speed pinion of the (elevation) reduction gear with the same gear ratio as for azimuth movement.

The unit is factory assembled and checked and the encoders are set. Then the mechanism is transported to the site, fitted with the reflector and fixed on top of the post. No adjustment is necessary on site.

The CETHEL drive unit is supported on an iron seating crown placed on top of the reinforced concrete post (height $\pm$ 2 cm, level $\pm$ 0.5%, south-reference aligned $\pm$ 1°, verified by water level and theodolite).

The foundation and post and their reinforcements were designed to satisfy Italian and French regulations for wind pressures, etc., with the reflecting panel in the most unfavourable loading conditions.

After completion of heliostat installation the sun image given by each of the CETHEL units was checked on the auxiliary target placed under the receiver. The focussing of a few of them had to be improved because of a lack of accuracy or perturbations during adjustment.

MBB drive units, columns and foundations

The main characteristics of the MBB drive assembly are:

motors	24 V DC, 300 rpm (3000 rpm for fast manoeuvres)
gears	5 stage spur gear, 1 stage worm gear, final spur gear

reduction ratio	Az 225 380, El 171 900
drive speed	Az 0.48°/min and 4.79°/min; El 0.63°/min and 6.28°/min
max. allowable mirror torque:	Az = 3330 Nm; El = 4170 Nm
total gear efficiency	0.35
bearings	preloaded ball bearings
position feedback	13 bit digital encoders
pointing errors in 18 km/hr wind (1 sigma):	
tracking	0.6 m rad
adjustment	0.3 m rad
deformation of pedestal and gear drive:	0.25 m rad
Total error	0.7 m rad

Since the determination of the actual attitude has a very large impact on the overall tracking accuracy, MBB fits 13-bit absolute value angular shaft encoders to the output axes of the gears thereby compensating for backlash, etc. The resolution of the 13 bit encoder is 0.8 mrad.

The drive unit itself (Fig. 3.1.8 in its final assembly stage) is covered by a glass fibre cover with access-holes for maintenance operations.

MBB's steel column, bolted to the concrete foundation (Fig. 3.1.9 shows the mould), supports the drive assembly consisting of the bearings, gears, motors and digital encoders (13-bit for 360° which give absolute angular information, derived directly from the moving axis).

The design of the MBB heliostat was experimentally verified in 1978 by means of a prototype and also by using the first heliostat manufactured according to the final manufacturing drawings. The latter, particularly, confirmed the performance characteristics, given above, including the accuracy of the reflected beam, which are well within specifications.

3.1.3 Heliostat electronics

CETHEL and MBB have both installed electronics in their heliostats but MBB has elected to give individual heliostats more autonomy - each "knowing" its own position relative to the tower and target and "knowing" the direction to the sun. CETHEL has a higher data-rate communication system, the Central Unit giving such data to the heliostats directly.

CETHEL heliostat electronics

A microprocessor-based local electronic system in each heliostat takes care of information exchange with the central unit and of the functions of heliostat command and control (Fig. 3.1.10). Particular emphasis has been put on failure-detection. The data transmission lines between the CETHEL central unit and heliostats are of multidrop serial, asynchronous, current loop type, with a modulation rate of 2400 bauds.

The control logic of each heliostat is able to receive from the central unit the tracking position values for azimuth and elevation every six seconds, compare these values with the current values of the encoders and decide

Figure 3.1.8 - MBB Drive Unit (Cover Removed)

Figure 3.1.9 - Mould for MBB Foundation

A. GENERAL CONCEPT

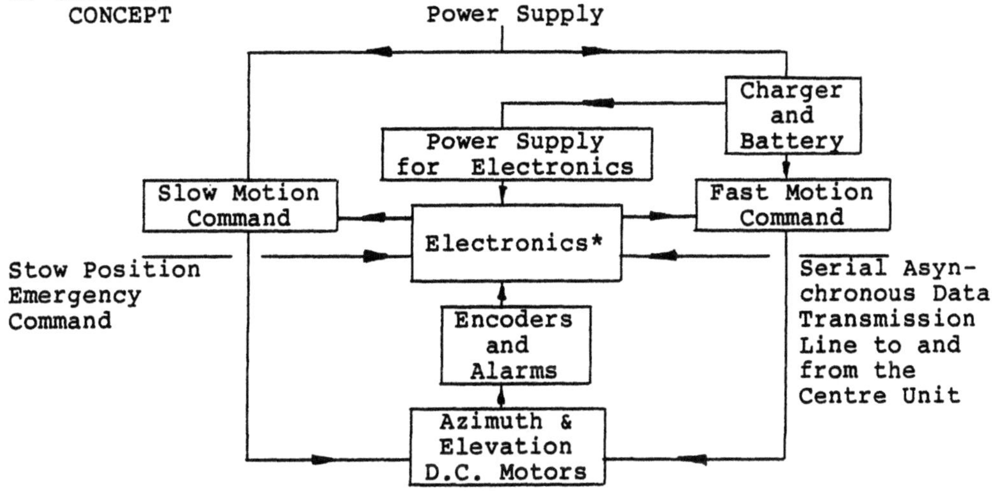

*B. DETAIL OF ELECTRONICS

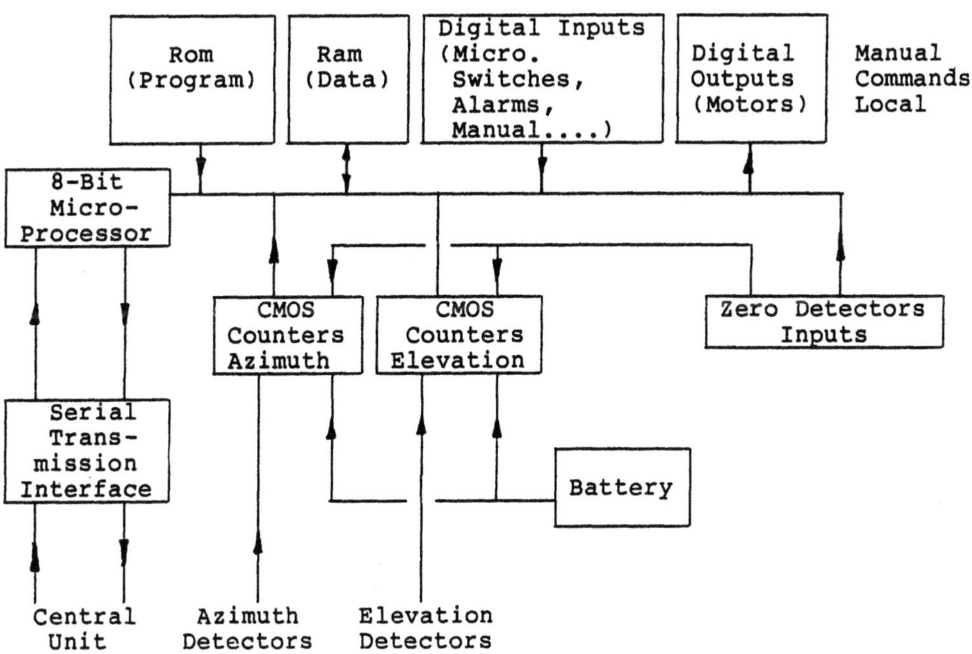

Figure 3.1.10 - The CETHEL Control System at Heliostat Level

at which speed and in which direction (sense) the heliostats must move. It also controls the alarms. The logic is primarily CMOS and TTL LS technology and, except for the DC motors and the encoders, the system is all mounted in one box.

Start-up

Once the current is established in the grid, the logic starts immediately. It checks whether the 24 V battery is charged. If not the logic informs the central unit and forbids fast speed.
If the battery is charged, the logic commands the closure of a relay, allowing operations with the battery.
The logic checks the state of the battery supplying power to the detectors, to inform the central unit of information loss. The counters can be reset to zero by a manual procedure.

Normal operation

The logic receives the commands from the central unit and commands the rotation of the motors to reach the assigned position. It also has a supervision role of heliostat operation.
The central unit may send an assigned position or ask for the current state or current position of the heliostat, automatic zero positioning and automatic command of speed. Other commands may be envisaged.
The logic compares the current position with the assigned positions in order to command both the sense and speed of rotation of the motors. A change of direction is always preceded by a delay to minimise mechanical stress.
The counters are forced to zero when the corresponding detectors are active.
When a movement takes place out of the normal range of operation, the logic cuts the power supply of the corresponding motor.

Emergency-command-to-stow-position transmission line

This line is shared among the heliostats of a group.
In case of failure of the field power supply, the operator can still order the heliostats to go to stow position, by depressing a button on the central console.
In this case the heliostats can move from the receiver position to stow position, following safety trajectories consisting of 12 points whose azimuth and elevation are computed for each heliostat, by the central unit, and sent every 72 seconds to the heliostats and memorized.

Central-unit-to-heliostat transmission line

This is a serial asynchronous current loop operating at 1200 baud. To provide failure detection current flows when there is no message. Data is transmitted as 8-bit words plus parity and stop bits:

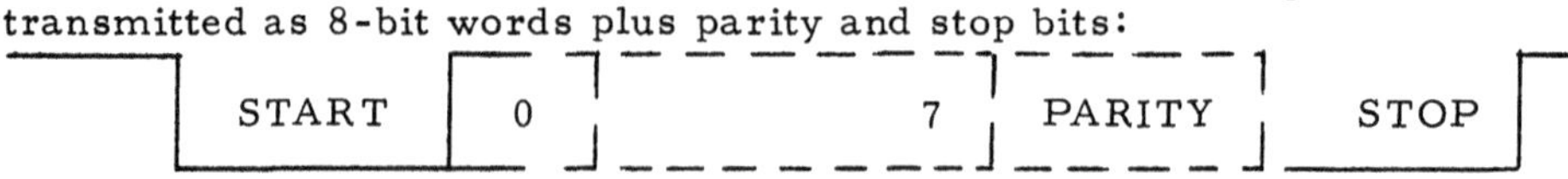

The line is organised as a bus, the failure of one heliostat not creating any disturbance to the rest of the line.
The line is optically isolated from the heliostat.

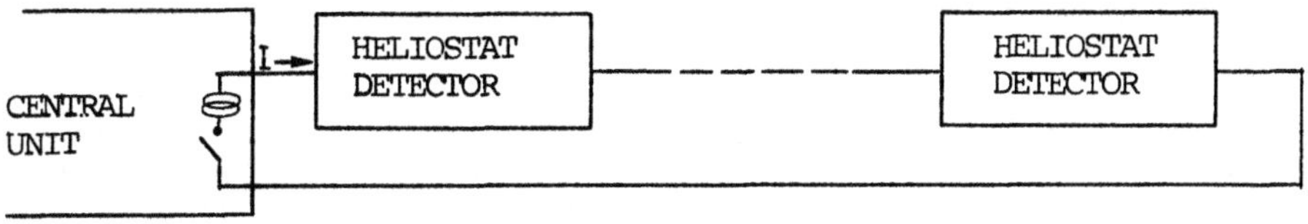

Heliostat-to-central unit transmission line

The general characteristics of this bus are similar to the above except for the parallel rather than series configuration.

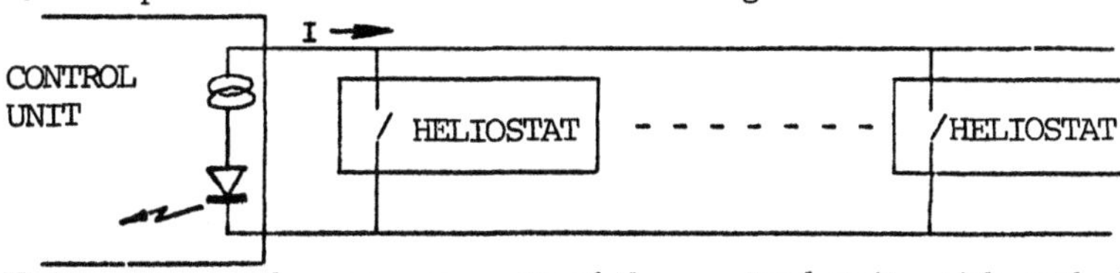

Heliostats send, upon request of the central unit, either their azimuth and elevation positions, or a word representative of their logical status. This status word is coded as one or two 8-bit words and contains information relative to:

- manual or automatic operation mode;
- alarms (general power supply);
- alarms (local power supply);
- alarms (batteries);
- micro-switches for extreme positions actuated or not;
- alarms (lack of rotation).

MBB heliostat electronics

The MBB tracking concept, which has been developed since 1978, provides somewhat more autonomy to the individual heliostats than does CETHEL's. Each heliostat can automatically track and maintain:

- stowage position (mirrors face down);
- standby position (near the receiver);
- receiver position;
- alignment position (to the auxiliary target).

The reflected beam is also controlled during the movement from one automatic position to another, to stay on the defined safety trajectories. Single heliostats and groups of heliostats can be commanded by the CU operator to manually defined positions for maintenance, washing and repair.

Figure 3.1.11 shows a simplified block diagram of the MBB tracking/servosystem. Fig. 3.1.12(a) illustrates the hardware related to every heliostat. Each local heliostat unit consists of an 8 bit microprocessor including program storage, battery buffered random access memory, timers, serial and parallel input/output circuits and in addition to a "simple" system, a separate arithmetic device. It performs at the same time input and decoding of commands, calculation of the sun position, determination of the real axis positions and servo-amplifier control and it sends status or position information to the Central Unit (C) upon request, by executing a multitasking program.

The heliostat control electronics receive serial information (commanded image position, date and time, data request) via the Command Data Bus from the MBB Central Unit in the plant Control Room. A single and a group address is dedicated to each heliostat in the field. The heliostat electronics decodes this information. It calculates the sun position and the related positions of the two axes more than 10 times every second. It compares the calculated positions with the actual positions retrieved directly from the angular encoders and sends drive signals to the servo amplifiers which power the motors until the right positions are reached. It sends, if requested, status and position informations back to the CU via the serial Response Data Bus.

The signals on the Command Data Bus and the Response Data Bus are in accordance with CCITT V. 24 (DIN 66 020) standard. The baudrate is 110 bit/s.

In addition two separate signal lines are wired from the CU to each group of heliostat electronics. These signal lines are used only in case of emergency (e.g. serial bus breakdown, emergency in the thermal system). They excite directly the two motor-drive electronics without involving the microcomputer. One signal (E1) turns the azimuth axis eastward with maximum speed. With the second (E2) the elevation axis is turned down to the stowage position. The motor drive electronics convert and combine the signals from the computer, the emergency lines, and the end position switches to power signals by providing to the DC-motors a high (fast) or a low (slow) voltage in the right direction.

The installed power supply converts the 220 V into the voltages needed by the microcomputer, the angular encoders and the drive electronics. It is a high efficiency primary-switching power converter. A small battery in each heliostat protects against loss of data in RAM in the event of a short loss of power.

Special purpose test units (Fig. 3. 1. 12(b)) have been provided to assist with the installation and maintenance of the MBB heliostats.

3.2 RECEIVER

The receiver is a once-through circular cavity-type steam generator based on the design concepts of the late Prof. Francia of Genoa. It is installed at the top of a tower at a height of 55 m, the centre line being inclined downwards 22° from horizontal (Fig. 3. 2. 1) towards the heliostat field.

The main waterway within the receiver consists of two virtually identical, parallel and independent branches, each 482 m long. This pipework is divided into four sections (Figs. 3. 2. 2 to 3. 2. 4).

i) an economiser (ECO), an open bottomless basket-like structure in the centre of the cavity designed to preheat the incoming feedwater to about 200 - 280°C;

ii) a truncated conical evaporator section (EVP), forming the greater part of the wall of the receiver and designed to boil the feedwater

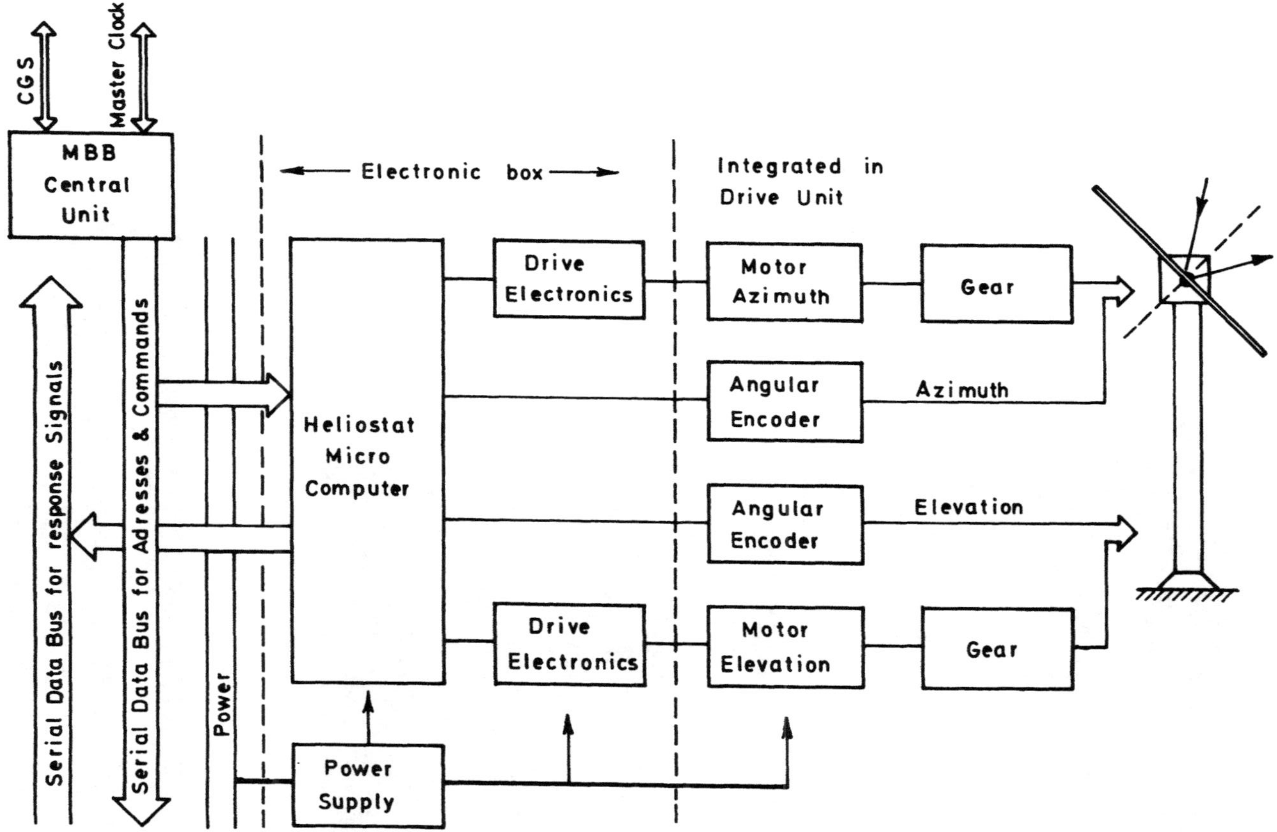

MBB Tracking System Heliostat for EURELIOS

Figure 3.1.11 – MBB Tracking Servo System

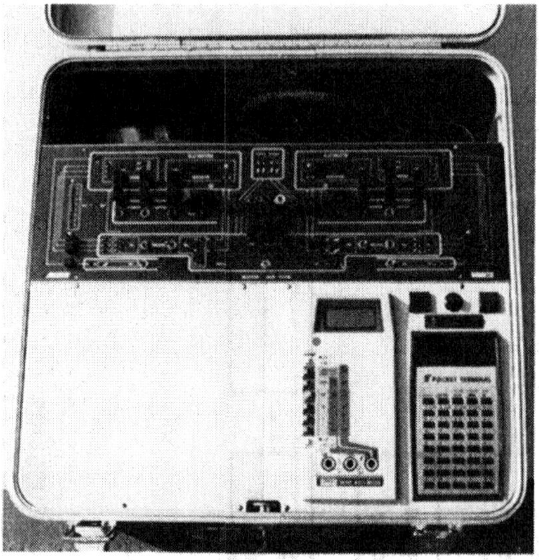

Heliostat Unit Tester for Servo System (HUT)

MBB Electronic Box

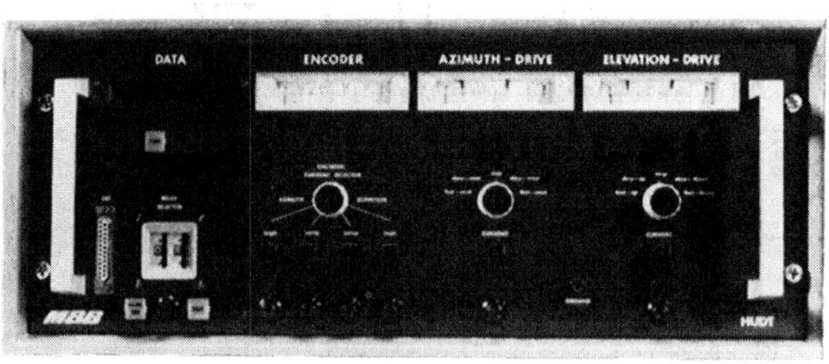

Heliostat Unit Drive Tester for Drive Unit Testing (HUDT)

Figure 3.1.12 - MBB Heliostat Electronic Units

Figure 3.2.1 - The Receiver and Tower

with very little superheating;

iii) a shorter, wider angle truncated conical primary superheater (SH1), within which the steam is superheated to between about 350 and 400°C. It is mounted immediately above the evaporator;

iv) a cylindrical secondary superheater (SH2) consisting of interlaced circular and straight pipes forming an open square-cell honeycombe-like structure in which the steam is brought up to its final state (nominally 512°C at a pressure of 62 kg/cm^2).

The receiver pipework is of A213 TP321 stainless steel painted with matt black Pyromark 2500 paint to increase its absorbtivity. The pipe diameter is 44.5 mm in the economiser (where the fluid is liquid) and 57 mm thereafter, the total length being 835 m. The two branches are fed separately, the main feedwater pipe having bifurcated prior to the flow meters and control valves which are on the access gantry at the base of the tower. The branches come together again into a common "main steam" pipe at the end of the secondary superheater, immediately before leaving the receiver. The receiver, without auxiliaries (valves, support structure, etc.) and lagging, has a volume of 3.4 m^3 and weighs 11.57 tons.

The overall concept of Prof. Francia's receiver includes the use of Pyrex glass sheets to improve its efficiency.

For EURELIOS these are to be installed inside the cavity, as shown by the distorted M-shaped section in Fig. 3. 2. 2. There they will offer a "greenhouse effect" for the hotter parts of the pipework, leaving the cooler economiser (ECO) open to direct illumination by the heliostats. The Pyrex will not be installed until sufficient experience has been gained of routine operation without it to permit a detailed analysis of its quantitative benefits.

To minimise heat loss, the outside of the receiver is covered with a mineral wool blanket (?packaged in glass-fibre sacks?) attached to an open-mesh stainless steel net and enclosed within a riveted stainless steel casing which retains and protects the insulation from the weather. The individual elements of the casing are preformed to make fins (Fig. 3. 2. 5) designed to minimise the adverse effects of possibly maladjusted heliostats focussing on the outside of the receiver. For the same reason, and also to protect the receiver in the event of a total electrical power failure which might cause all or part of the spot from the heliostat field to drift slowly to the east as the sun moved west, the rim is protected by a circle of silicon thermal protection bricks (Fig. 3. 2. 6).

The receiver includes appropriate monitoring and control equipment connected into the control room. The flow conditions under typical circumstances with a steam flow rate of 4200 kg/h at 62 kg/cm^2 and 510°C, result in a transit time for water (from the ECO inlet to the superheater outlet) of about 16 minutes. This implies that careful design of monitoring and control systems is essential to maintain steady operating conditions despite the comparatively rapid variations of incoming flux which result from the frequent presence of clouds.

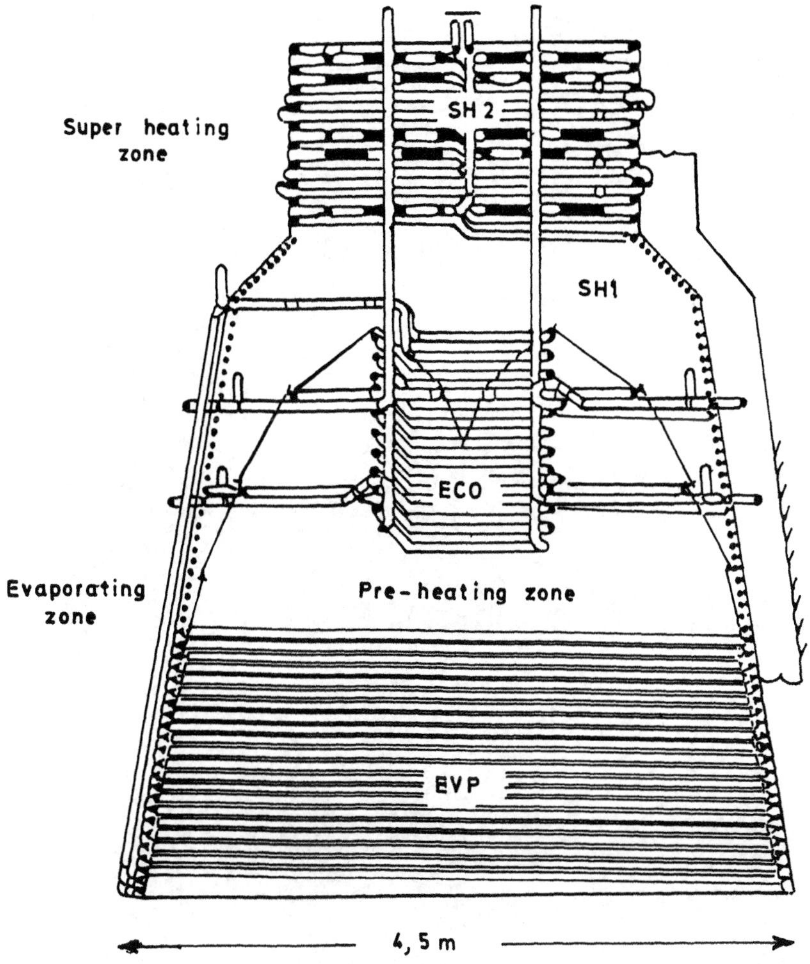

EVP = Evaporating section
ECO = Economizer
SH1 = Primary superheater
SH2 = Secondary superheater

Figure 3.2.2 - Receiver - A Sectional View

Figure 3.2.3 - View into the Mouth of the Receiver

Figure 3.2.4 - The Receiver and Its Supporting Structure (Without Auxiliaries or Lagging)

Figure 3.2.5 - The Receiver Prior to Hoisting

Figure 3.2.6 - The Receiver at the Top of the Tower (Showing Auxiliary Target and the Fitting of the Attemperator Access Gantry)

First, a safety valve at the receiver outlet, set at 64 kg/cm^2, protects against overpressures. Second, the receiver contains 80 thermocouples and 47 flux meters welded to the tubing. The average tube temperature and average heat flux into the tubes obtained from these sensors are used by the control system to regulate, instant by instant, the feedwater flow rate. The receiver also contains a number of water temperature and pressure sensors which are used by the regulation system, particularly to control relative to an operator-selectable secondary-superheater temperature set point.

Finally, the receiver is provided with a desuperheating/attemperating system to keep the output steam temperature as constant as possible as the solar flux varies, particularly to avoid temperature increases above the design figures. This system consists of four automatically controlled Yarway sleeve-type attemperator spray valves, two in each branch. The first pair, one at the end of each evaporator section, is controlled on the basis of superheater inlet temperature and can provide up to 8% of the maximum total flow. The other pair, at the inlet to the secondary superheater, controls superheater outlet temperature introducing up to 4% of the maximum total flow. The attemperators are all fed from a third branch of the feedwater piping. The pressure of this branch is controlled at a differential of about 12 kg/cm^2 over the pressure in the two main branches so as to guarantee that the water remains liquid and flows fast enough. The water temperature is raised to about 190°C in a heat exchanger situated between the economiser (ECO) and the evaporator, so as to minimise thermal shocks at the attemperators themselves.

The receiver is also protected by the monitoring of a number of other parameters. On the basis of these, the heliostat field is moved out of the receiver in the event of:

. a very low feedwater pump delivery pressure;
. a high differential between the flow rates in the two main branches;
. a low feedwater flowrate at the A/B branch inlet;
. the loss of either the 110 V or the 380 V power supply.

Further, in the event of loss of the 380 V AC supply and consequent loss of power to the feedwater pumps and the heliostat subsystems, when the basic feedwater pressure has dropped from the normal 101 kg/cm^2 to 86 kg/cm^2, a nitrogen-pressurised auxiliary water supply (Section 3. 3. 2) automatically continues to feed the receiver until the solar spot has moved clear of the receiver due to the east-to-west motion of the sun.

Structurally the main strength of the receiver is in the pipework itself. It is supported, along four generators of its conical shape, from a pair of substantial mild steel three-quarter-circular tubes (visible in the figures). The whole receiver is mounted on a trolley which can be winched to the top of the tower (Fig. 3. 2. 7) on integral rails. There the final connections are made to the steam cycle, control and monitoring systems.

Figure 3.2.7 - The Receiver Being Winched up the Tower

3.3 STEAM CYCLE

3.3.1 Description

The steam cycle (Fig. 3.3.1) is a conventional single superheating Rankine Cycle consisting of a boiler (the receiver, Section 3.2), a steam turbine (Section 3.4) and a condenser, with appropriate pumps, valves, controls, safety systems, etc. Associated with it is an auxiliary system designed to keep the plant operating for up to about 30 minutes in the event of cloud passing, using saturated steam from an accumulator (storage tank) superheated in a heat exchanger by a second working fluid, a hot molten salt (Section 3.5).

Typical operating parameters are given in Table 3.3.1. Starting from a temperature of 25°C it takes about 45 minutes to heat the receiver to 350°C, the temperature for turbine start-up, under conditions of full insolation - somewhat longer under normal early morning conditions. During this period the water or steam produced by the receiver is taken through a bypass valve (N11-F301) and attemperator/desuperheater into a Flash tank (N11-A002) before being passed into the Condenser (N61-3001) and pumped around the system again. The control system (Section 3.7.2) maintains the upstream pressure by adjusting the bypass valve. The control system also automatically adjusts the water level and pressure in the flash tank by means of the two valves N11-F309 and N11-F308 which, respectively, allow water and steam to flow to the condenser.

The condenser itself is an evacuated four-pass water cooled unit with a nominal operating pressure of 6 cm Hg (reduced to 3 cm Hg when using the salt storage system).

Once the steam temperature is high enough, steam can be admitted to the turbine, first via valve N11-F307 and then by the main steam control valve N11-F303 (the rotor having been turned over by the starter motor and the steam seals fed from the flash tank). This results in automatic restriction of the bypass valve to maintain the steam pressure. The (low pressure) exhaust of the turbine is passed directly to the condenser.

The 7-stage steam turbine runs at a speed of 8195 rpm, corresponding to 1500 rpm at the generator (alternator) after the appropriate reduction gearing, when synchronised with the 50 Hz main grid. The turbine is regulated by a Woodward regulator acting on the nozzle valves.

Condensate in the well of the condenser is extracted by two centrifugal "booster" pumps, N21-C001 A/B, each designed to extract 8800 l/hr when the condenser is evacuated to a pressure of 3 cm Hg. The water level in the condenser is controlled by the regulation system. This allows additional demineralised water to enter from (or causes excess to be pumped by the booster pumps back to) the cold water storage tank. This is particularly important during start-up of the plant from cold: water levels in the cycle tend to rise due to the conversion of water to steam in what was initially full of water. Air and uncondensable vapours are extracted from the condenser by two vacuum pumps, normally one in service, one in standby.

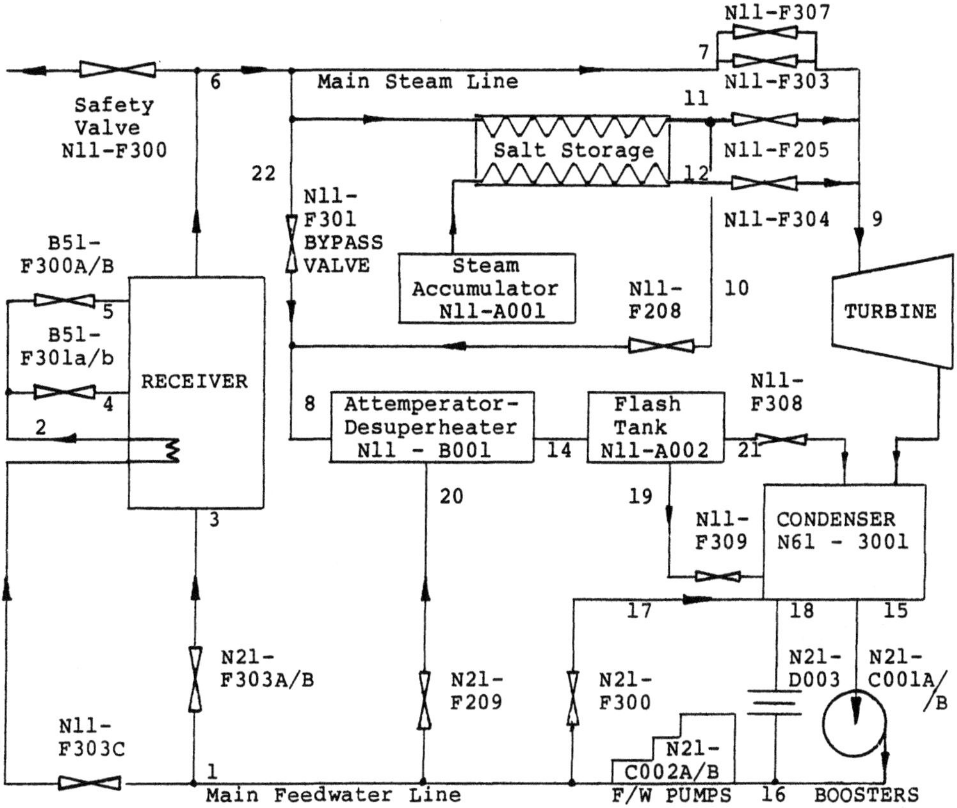

The numbered 'points' of Table 3.3-1 are shown on this sketch. Flow rates for A or B valves etc. are half the total flow rate in the common line.

DEFINITION OF MODES OF TABLE 3.3-1

- A Pressurising main feedwater line
- B Filling receiver
- C Receiver minimum load, on turbine
- D Receiver maximum load, on turbine
- E Receiver maximum load, 100% on bypass
- F Beginning of Salt Storage system intervention
- G End of storage system intervention (receiver isolated)
- H Receiver minimum load, heating salt system into bypass
- I Receiver maximum load, heating salt system into turbine.

Figure 3.3.1 - Steam cycle

MODE A																						
Point	1	2	3	4	5	6	7	8	9	10	11	12	13	14	15	16	17	18	19	20	21	22
Q=kg/h	0	–	0	–	–	0	–	–	–	–	–	–	–	0	8500	8500	8000	500	–	–	–	–
P=ata	–	–	–	–	–	1	–	–	–	–	–	–	–	1	1	2.2	1	1	–	–	–	–
T=°C	–	–	–	–	–	24	–	–	–	–	–	–	–	24	24	24	24	24	–	–	–	–

MODE B																						
Point	1	2	3	4	5	6	7	8	9	10	11	12	13	14	15	16	17	18	19	20	21	22
Q=kg/h	1000	–	1000	–	–	1000	–	–	–	–	–	–	–	1000	8500	8500	7000	500	1000	–	–	1000
P=ata	101	–	46	–	–	40	–	–	–	–	–	–	–	1	1	2.2	1	1.9	1	–	–	46
T=°C	25	–	25	–	–	25	–	–	–	–	–	–	–	25	25	25	25	25	25	–	–	25

MODE C																						
Point	1	2	3	4	5	6	7	8	9	10	11	12	13	14	15	16	17	18	19	20	21	22
Q=kg/h	1000	120	446	80	40	1000	–	–	1000	–	–	–	–	–	8500	8500	6900	600	–	–	–	1000
P=ata	101	75	72	66	65	64	–	–	12	–	–	–	–	–	.06	1.8	.06	.06	–	–	–	64
T=°C	25	150	25	150	150	510	–	–	500	–	–	–	–	–	25	25	25	25	–	–	–	510

MODE D																						
Point	1	2	3	4	5	6	7	8	9	10	11	12	13	14	15	16	17	18	19	20	21	22
Q=kg/h	5000	600	4400	400	200	5000	5000	–	5000	–	–	–	–	–	8500	8500	2900	600	–	–	–	–
P=ata	101	76	76	67	64	64	63	–	61	–	–	–	–	–	.06	1.2		.06	–	–	–	–
T=°C	36	180	36	180	180	510	510	–	510	–	–	–	–	–	36	36	36	36	–	–	–	–

MODE E																						
Point	1	2	3	4	5	6	7	8	9	10	11	12	13	14	15	16	17	18	19	20	21	22
Q=kg/h	5000	600	4400	400	200	5040	–	–	–	–	–	–	–	6400	8500	8500	1500	600	3200	1400	5000	5000
P=ata	101	76	76	67	64	64	–	–	–	–	–	–	–	50	.06	1.9	–	1.9	5	5	5	6.3
T=°C	40	180	40	180	180	510	–	–	–	–	–	–	–	160	40	40	40	40	150	40	150	510

MODE F																						
Point	1	2	3	4	5	6	7	8	9	10	11	12	13	14	15	16	17	18	19	20	21	22
Q=kg/h	1000	–	1000	–	–	1000	–	–	750	–	–	750	750	8400	8500	8500	6900	600	1200	1400	120	1000
P=ata	101	–	36	–	–	64	–	–	8	–	–	17.9	19	5	.06	1.9	.06	1.9	5	5	5	64
T=°C	25	–	25	–	–	510	–	–	440	–	–	440	210	160	25	25	25	25	150	25	150	510

MODE G																						
Point	1	2	3	4	5	6	7	8	9	10	11	12	13	14	15	16	17	18	19	20	21	22
Q=kg/h	–	–	–	–	–	–	–	–	300	–	–	300	300	–	8500	8500	7900	600	–	–	–	–
P=ata	–	–	–	–	–	–	–	–	7	–	–	8.6	9	–	.03	1.8	–	1.8	–	–	–	–
T=°C	–	–	–	–	–	–	–	–	400	–	–	440	175	–	25	25	25	25	–	–	–	–

MODE H																						
Point	1	2	3	4	5	6	7	8	9	10	11	12	13	14	15	16	17	18	19	20	21	22
Q=kg/h	2500	300	2200	200	100	2500	1000	1500	1000	1500	1500	–	–	–	8500	8500	4000	600	–	1400	–	–
P=ata	101	76	73	65	64	64	63	8	12	5	5	–	–	–	.06	1.8	.06	1.8	–	5	–	–
T=°C	28	150	28	150	150	480	480	430	480	310	330	–	–	–	28	28	28	28	–	28		

MODE I																						
Point	1	2	3	4	5	6	7	8	9	10	11	12	13	14	15	16	17	18	19	20	21	22
Q=kg/h	5000	600	4400	300	200	5000	3500	–	5000	–	1500	–	–	–	8500	8500	2900	600	–	–	–	–
P=ata	101	76	76	67	64	64	63	–	61	–	62	–	–	–	.06	1.9	.06	1.9	–	–	–	–
T=°C	36	150	36	150	150	480	480	–	450	–	430	–	–	–	36	36	36	36	–	–	–	–

NOTE : See figure 3.3.1 for Definition of modes and points

Table 3.3.1 - Steam Cycle Design Values

The output of the centrifugal booster pump(s) feeds either of the standby-redundant pair of reciprocating "feedwater" pumps N21-C002 A/B. Each of these delivers 8200 l/hr to the main feedwater line at the normal operating pressure of 101 kg/cm^2.

As mentioned in the description of the receiver, the flow of feedwater to the receiver is controlled (by valves N11-303 A/B/C) so as to regulate the temperatures achieved (subject always to the pressure control achieved by the steam cycle's turbine-bypass valve N11-F301). Any excess water, above that needed to maintain the normal operating feedwater pressure, is returned to the condenser through a "feedwater pressure control" valve N21-F300.

Water supply to the bypass desuperheater and to the storage steam accumulator is also taken directly from the main feedwater line. Water hammer in the feedwater line is damped by a bank of four tanks containing pre-pressurised (70 kg/cm^2) nitrogen bags and located just downstream of the reciprocating feedwater pumps.

The steam cycle's remotely operated valves are all electrically controlled and pneumatically operated. Compressed air is supplied by two two-stage reciprocating compressors controlled to give a working pressure of 75 kg/cm^2 in the instrument air circuit. The air from the compressors is cooled by the turbine oil cooler circuit: a closed cycle of demineralised cooling water driven by a pair of centrifugal pumps whose main purpose is to keep the turbine oil below 38^oC at all times.

The last, but not least important, function of the steam cycle is to provide both superheated and saturated steam to the salt-storage system (Section 3. 5) and to utilise the condensate or steam from it. To charge (i. e. heat) the salt, main steam (from the receiver) is admitted by valve N11-F203, the flow being controlled by the salt storage system's own controls (nominal flow rate 1010 kg/h) and finally passed either through N11-F208 to the bypass desuperheater or optionally through N11-F205 to be mixed with the main steam supply to the turbine. To charge the steam accumulator, N11-A001, it is first partially filled with water from the main feedwater, condensing and raising both pressure and temperature until the accumulator of saturated steam and water reaches its nominal 19 kg/cm^2, 210^oC charged state.

Once the salt and the steam accumulator are charged and the system enabled, the steam cycle control will automatically transfer to use the stored steam and energy to keep the generator running for up to 30 minutes whenever a cloud interrupts insolation. This is implemented by starting the transfer sequence as soon as the pressure of the main steam falls below 8 kg/cm^2. Similarly, when the cloud has passed and the steam quality is acceptable the turbine will again automatically be fed from the main steam line. The system has been designed with flexibility in mind and the optimising of the detailed operating procedures will be an interesting part of the Phase D research programme.

3.3.2 Emergency feedwater system

This system (Fig. 3.3.2) is used only in case of emergency. It ensures that if the pressure in the feedwater line drops below 86 kg/cm^2 due to:

- loss of power to the booster or feedwater pumps, or
- electrical or mechanical failure in the feedwater system,

sufficient water is supplied to the receiver to protect it whilst it is still illuminated, namely for the roughly 20 minutes required for the motion of the sun to take the spot from the stationary MBB heliostat field out of the receiver.

The system consists of 6 one cubic metre bottles with a design pressure of 220 atm. Three of these bottles (N21-A003A/B/C) contain nitrogen at 200 atm., the others (N21-A002A/B/C) demineralised water for the receiver. A bank of six nitrogen storage bottles (N21-A004A/B/C/D/E/F) make up gas losses, if any. The compressor N21-C005 reinstates the system after use by transferring nitrogen back from the three water bottles to the main nitrogen bottles (N21-A003A/B/C) thereby allowing the water bottles to be refilled from the main feedwater line.

3.3.3 Steam cycle alarms

The steam cycle includes alarms to warn the operation of the following abnormalities:

- steam storage tank (N11-A001)
  - high condensate level 1600 mm
  - low condensate level 1100 mm
- condenser (N61 3001)
  - condensate low level 200 mm
  - condensate very low level 100 mm
  - condensate high level 600 mm
  - condensate very high level 720 mm
- flash tank (N11 A002)
  - high level 400 mm
  - low level 100 mm

3.4 TURBINE

The turbine is a 7-stage impulse turbine rigidly coupled to the reduction gear pinion. The reduction gear is connected to the alternator through a flexible joint. The combined turbo-generator unit, mounted on its chassis is illustrated in Fig. 3.4.1. General data are given in Table 3.4.1 and Table 3.4.2 gives nominal heat balance calculations for five sample turbine operating states.

The turbine casing is cast in two halves in a steel alloy. The top of the casing houses the nozzle diaphragms in grooves provided with drain holes. Leakage of steam between stages is minimised by metal labyrinth seals and there is a steam sealing box on the steam inlet and the exhaust side. The rotor is a special steel forging, heat treated to give the necessary high resistance to mechanical and thermal stress.

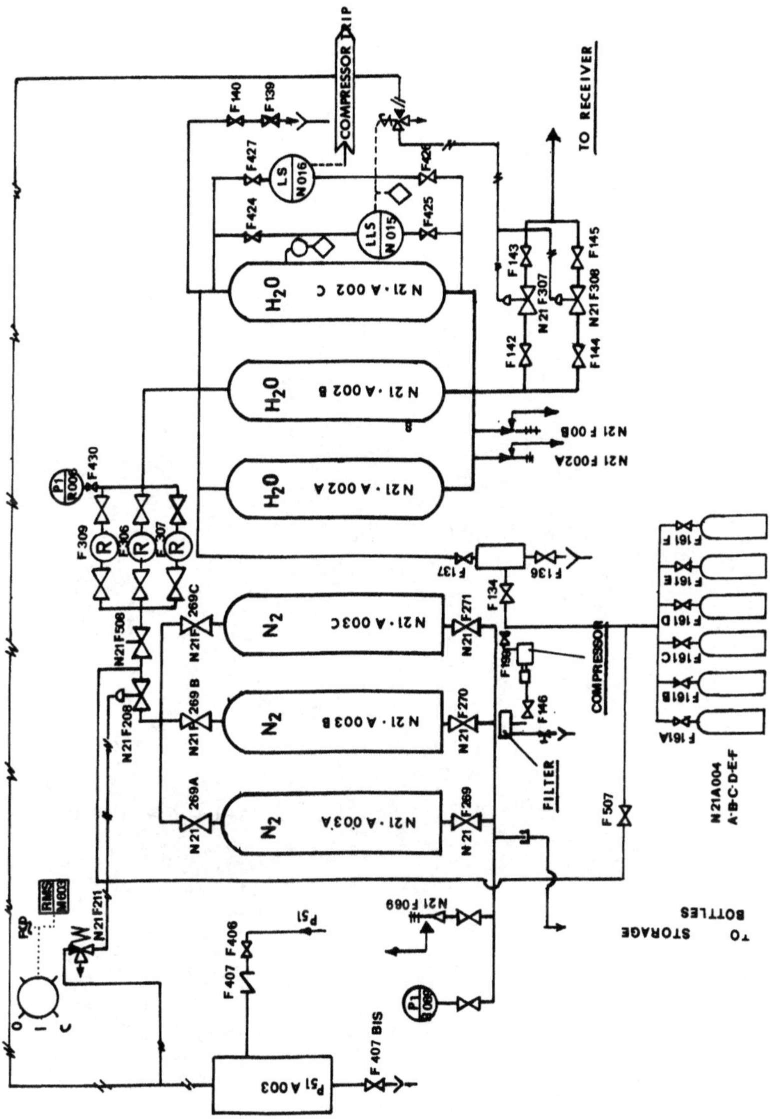

Figure 3. 3. 2 - Emergency Feed Water System

Figure 3.4.1 - The Combined Turbo-Generator Unit

TURBINE

- PART NUMBER	No. 1953
- MANUFACTURERS	ANSALDO - GENOA
- JOB	No. 18.477
- CONSTRUCTION YEAR	1980

STEAM CONDITIONS AT THE STEAM VALVE (Turbine design point)

- PRESSURE	60 ata
- TEMPERATURE	510°C
- FLOWRATE	5200 Kg/h

STEAM CONDITIONS AT THE TURBINE EXHAUST

- PRESSURE	0.06 ata
- TEMPERATURE	36°C

OPERATING DATA

- OUTPUT AT THE TURBINE/GENERATOR JOINT	1200 kW
- TURBINE SPEED	8195 rpm
- GENERATOR SPEED	1500 rpm
- OVERSPEED EMERGENCY TRIP	9250 rpm

OVERALL DIMENSIONS & WEIGHT OF THE COMPLETE TURBO-GENERATOR SET

- TOTAL LENGTH OF THE SET (TURBINE, REDUCTION GEAR, GENERATOR)	5.465 m
- MAXIMUM WIDTH	2.160 m
- MAXIMUM HEIGHT	2.400 m
- WEIGHT OF TURBINE AND ACCESSORIES	3746 kg
- WEIGHT OF ALTERNATOR (Power generator)	5000 kg
- TOTAL WEIGHT OF THE SET & ACCESSORIES	13500 kg

Table 3.4.1 - General Data on the Turbine

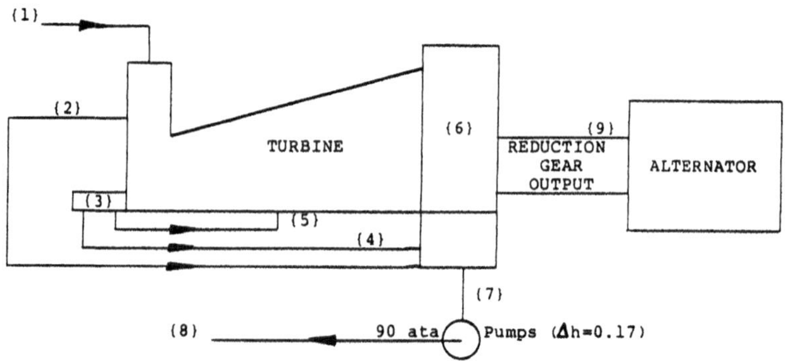

STATE		100% OUTPUT	** NOMINAL OUTPUT	75% OUTPUT	50% OUTPUT	AT START OF SALT STORAGE INTER-VENTION
POINT*	PARAMETER					
1. Input to Turbine	Flow, kg/hr	5200	5200	3900	2600	1445
	Pressure, ata	60	60	60	60	16
	Temp. °C	510	450	510	510	410
	Enthalpy, kCal/kg	823.11	788.89	823.11	823.11	782.22
2. Input Valve Leakage	Flow, kg/hr	60	60	60	60	16
3. Gland Seal Steam	Enthalpy, kCal/kg	782.58	750.46	778.19	773.93	745.20
4. Gland Seal Loss	Flow, kg/hr	42	41	26	15	11
5. Gland Seal Re-use	Flow, kg/hr	136	137	109	85	40
6. Turbine Exhaust	Flow, kg/hr	5098	5099	3814	2525	1418
	Enthalpy, kCal/kg	603.57	587.89	602.57	604.12	601.40
	Pressure, ata	0.06	0.06	0.048	0.038	0.033
7. Condenser Output	Enthalpy, kCal/kg	35.83	35.83	31.82	27.75	25.35
8. Feedwater	Flow, kg/hr	5200	5200	3900	2600	1445
	Enthalpy, kCal/kg	36	36	31.99	27.92	25.52
9. Reduction Gear Output	Shaft power, kW	1200	1100	894	570	210
GHR	kCal/kWh	3410.8	3559.1	3451	3627	5206.8
GSR	kg/kWh	4.33	4.72	4.36	4.56	6.88

* Point numbers correspond to those on the sketch above.
** This is an example of one combination of steam conditions which would give the nominal power output.

Table 3.4.2 - Examples of Turbine Heat Balance

3.5 THE MOLTEN SALT THERMAL STORAGE SYSTEM

3.5.1 General description

The purpose of the thermal storage system is to keep the turbo-alternator and the auxiliaries running at reduced power for up to 30 minutes without solar input, e.g. during cloud passing. It can also be used to warm the turbine and/or pipework (or to keep them warm for limited periods) if required.

The system consists of two main parts (see Fig. 3.5.1):

. an accumulator (a pressurised water tank) provided by ANSALDO, able to deliver saturated steam at a pressure decreasing from 19 bar to 7 bar, by auto evaporation;
. a salt storage subsystem, designed and implemented by CETHEL, to superheat the steam using as working fluid a molten salt, such as HITEC (an eutectic mixture: 53% KNO; 40% $NaNO_2$; 7% $NaNO_3$), and including two steam/salt heat exchangers and two salt storage tanks.

Energy is stored using the latent heat of this salt between 430 and 275°C, making it possible to superheat the saturated steam from the accumulator to a high enough temperature for the turbine (410°C). The nominal 60 kWh of stored energy corresponds to only about 1100 kg of salt.

The salt cycle concept adopted, namely to have two storage tanks, one hot and one "cold" (although still hot enough for the salt to remain molten), allows salt to be pumped at an essentially constant temperature into the salt/steam heat exchangers: from the cold tank at the cold tank temperature through the charging exchanger to the hot tank and vice versa for discharge when the salt is giving up heat to superheat steam from the accumulator to drive the turbine.

During charging some of the steam from the receiver is diverted through the first heat exchanger (flow rate 1010 kg/h, up to 64 bar, 480°C/510°C) where it heats salt being pumped from the cold to the hot tank.

When the salt is at 430°C in the hot tank, discharge can start if the steam pressure at the turbine inlet drops below the accumulator pressure, 19 bar (although discharge normally waits for the pressure to drop to 8 bar). The saturated steam from the accumulator then passes through the second heat exchanger, where it is superheated by the salt to about 410°C and delivered to the turbine. The salt is then being pumped back from the hot to the cold tank. The mean cold temperature of the salt is about 275°C.

The main components of the salt system are therefore:

- for the charging circuit:
  . a cold tank SF1 where the salt is stored at a temperature of 275°C;
  . a circulating vertical pump P1 installed upon this tank;
  . connecting pipes between P1 and the charging heat-exchanger EC-1;
  . the charging heat-exchanger EC-1;
  . connecting pipes between EC-1 and the hot salt tank.
- for the discharging circuit:
  . a hot tank SC1 containing the salt at a temperature of 430°C;

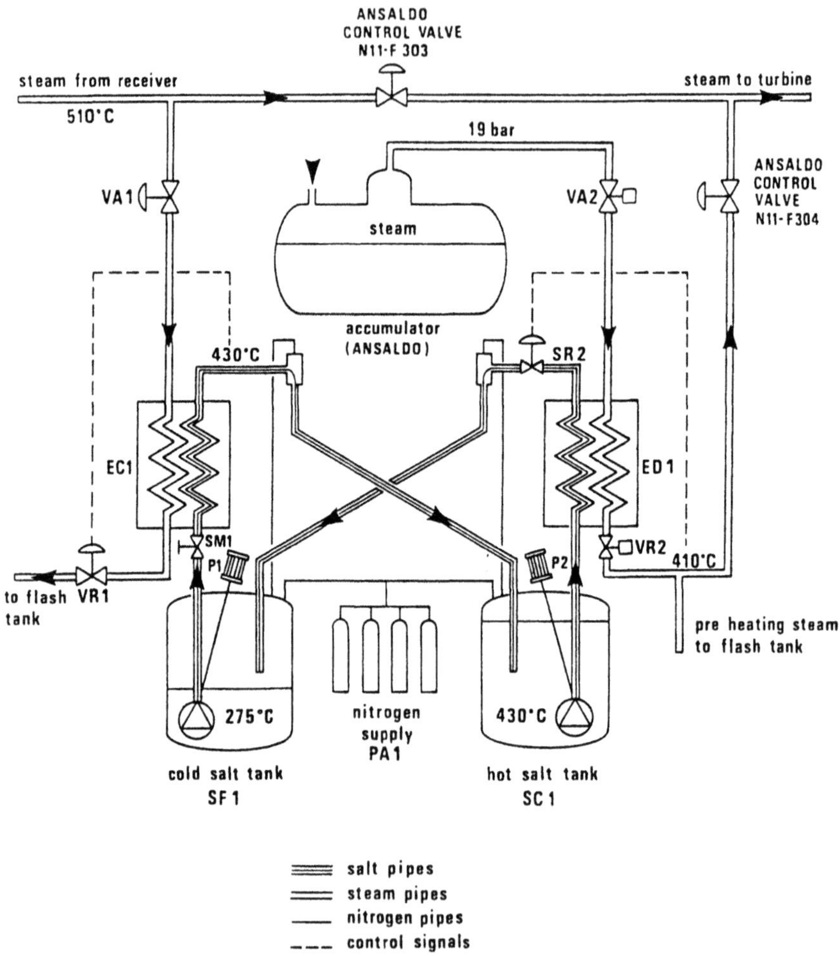

Figure 3.5.1 - Thermal Storage System (Main Components)

- a circulating vertical pump P2 installed upon this tank;
- connecting pipes between P2 and the discharging heat-exchanger ED-1;
- the discharge heat-exchanger ED-1 (a steam superheater);
- connecting pipes between ED-1 and the cold salt tank.

Each tank is large enough to hold the entire charge of salt, the ullage space being filled with nitrogen. All pipes, those for nitrogen as well as the salt circuit, are of stainless steel 304L, 1/2", schedule 10S.

The design of this system involved the solution of some interesting problems in Phases B and C but its operation on a day-to-day basis during the later phases is expected to provide even more useful insight into the strategy of using and/or keeping the stored energy in the design of larger thermal storage capacities.

The salt is of course solid at normal ambient temperatures (it freezes at about 160°C) and so the whole stainless steel circuit had to be designed to be self-draining into the tanks. The layout of the pumps and pipework in the tanks had to allow for electrical melting of the salt in such a way that its expansion prior to melting did not overstress the system and that there was no risk of starting a pump which still contained traces of solid salt. Further, to avoid any risk of salt freezing inside the pipes and heat exchangers and to prevent thermal shock when salt starts to pass through, these components have to be heated. They are kept at the correct temperature by means of electrical resistances, or tracing, and the whole inox steel salt circuit is self-draining (10% slope) into the tanks, in which the salt can freeze and be remelted by electrical heaters.

The design of heat-exchangers also had to fit with rather low heat-exchange rates and very small flows of both steam and salt, especially for the steam superheater. As a result, the heat-exchangers were made of concentric twin pipes throughout, the internal pipe being provided with external fins and partially filled with a core. In this way appropriately small cross-sections were achieved to provide acceptable heat transfer coefficients.

The compact layout of the installation and the overdesigned insulation are designed to result in reduced thermal losses. In addition two different electrical heating rates are available during day and night in order, by minimising it at night, to save auxiliary power. But during the day, if the system is to remain immediately operable at any moment, the temperature level of the salt and of the whole salt circuit, including empty sections, must be kept high enough to avoid excessive drops of temperature. Optimisation in this area will be one of the more valuable experimental activities with this subsystem during Phase D.
Additional numerical data are given in Table 3.5.1.

3.5.2 Operation and control

Charge

Valves VA1 and VR1 (refer to Fig. 3.5.1) are operated through electrically controlled pneumatic actuators.

HEAT EXCHANGERS	EC1. CHARGING HEAT EXCHANGER		ED1. DISCHARGE HEAT EXCHAN (SUPERHEATER)	
Power	110.8 kW		190 kW (78 kW)*	
Charging (discharging) time	40 minutes		30 minutes	
Heating surface	16 m^2		37.7 m^2	
Internal tube diameter	52.5 / 60.3 mm		42.7 / 48.3 mm	
External tube diameter	82.8 / 88.9 mm		82.8 / 88.9 mm	
Core	42.2 mm		33.4 mm	
Height of the fins	7.11 mm		12.7 mm	
	Between the two concentric tubes	Inside the internal tube	Inside the internal tube	Between th two concent. tubes
Fluid	SALT	STEAM	SALT	STEAM
Pressure (abs. bars)	-	64	-	19 (9)
Flow (kg/s)	0.455	0.280	0.937 (0.365)	0.401 (0.1
Entering temperature (°C)	275	480	430 (430)	210 (175
Leaving temperature	430	330	300 (293)	410 (410
Speed (m/s)	0.2	20.3	0.92 (0.4)	18.1 (10.
Pressure drop (bar)	0.1	1.4	1.1 (0.22)	1.1 (0.36

MOLTEN SALT TANKS

- Overall volume (each) — $1m^3$
- Diameter — 1.1m
- Overall height — 1.15m
- Salt capacity:
  - overall — 1,250 kg
  - for process — 1,100 kg

  (corresponding heat capacity 60 kWh)
- Temperature of the salt:
  - cold tank — 275°C
  - hot tank — 430°C
- Material — INOX 316L
- Insulation thickness — 200mm (Rockwool)

PIPING

Diameter 17.1/21.3 (½")

Material Stainless Steel 304L

ELECTRICAL HEATING

Installed Power 45 kW

* Minimum working conditions in parentheses.

Table 3.5.1 - Molten Salt Storage System Numerical Data

During charging the salt flow is approximately constant and the temperature of the salt leaving the heat exchanger is maintained around 430°C by varying the steam flow through the steam control valve VR1.

The nominal required temperature of steam entering the heat-exchanger is 480°C. A secondary control loop acts to anticipate opening of VR1 when the steam temperature becomes lower than 480°C and vice versa. VA1 is an on-off valve.

Discharge

In the steam side of the discharge circuit, VA2 and VR2 are on-off pneumatic valves. SR2 is a pneumatic control valve on the salt side. During operation the steam temperature leaving the exchanger is regulated around 410°C by varying the salt flow through SR2.

At the very beginning of operation, during the first 30 seconds about, SR2 is forced wide open in order quickly to fill the heat exchanger with hot salt for operation.

The steam flow is not controlled by the salt storage system, possibly decreasing with the pressure in the hot water accumulator, although the steam cycle control normally restricts the flow to maintain a pressure of 8 bar at the turbine inlet.

Controls

The salt storage is operated using three turn-push-buttons with signal lamps (TPL-1, TPL-2, TPL-3) located in the vertical panel in the Control Room. Buttons TPL-1 and TPL-2 control the charging circuit; TPL-3 controls the discharging circuit.

Depending on the positions of these buttons, the storage can be set in the following states:
. stopped for night;
. warming-up of the connecting pipes of the charging circuit;
. charging operation;
. warming-up of the connecting pipes of the discharging circuit, enabling discharge.
Two other switches (MHS and MCS) allow control of the electrical heaters in the salt tanks.

The salt storage panel in the control room also displays the temperatures of the salt in the hot and cold tanks and has clocks which measure the running times of the charge and discharge circuits, respectively.

On "run", turn-push-button TPL-1 allows warming of the charging circuit pipes to 275°C for the pipes between the cold tank and the charging heat-exchanger, EC1, to 430°C for the pipes between the heat-exchanger and the hot tank and also for the heat-exchanger itself. This operation takes place early in the morning, lasting about one hour, whilst the boiler is being started-up. When the boiler produces steam at nominal characteristics, button TPL-2 can be set to "run", to start the charging sequence. TPL-2 controls the startup of pump P1 and the electrical supply to the control system for valve VR-1, which is operated according to the temperature of salt leaving the heat-exchanger. Start-up is automatically forbidden if the temperature, in any section of the pipework, or the level of salt in the cold tank, is too low.

The charging operation can be stopped voluntarily by positioning TPL-2 on "stop", or automatically by the safety system. In the event of overpressure in the salt circuits (over 0.05 bar above atmospheric), which might be the result of a steam leakage into the salt, the safety system stops the pumps and closes the valves, isolating the steam circuits. The pressure measurement is performed in the nitrogen atmosphere maintained above the salt in the two tanks. This neutral atmosphere is in fact an extra precaution against oxidation of the salt, although it should already be stable enough at the envisaged temperature levels.

Turn-push-button TPL-3 on "run" causes warm-up of the discharging circuit pipework (to 430°C for the pipes between the hot tank and the heat-exchanger ED-1, to 275°C for the pipes between the heat-exchanger and the cold tank). After 1 hour the circuit is ready for discharge provided the salt is hot (430°C in the hot tank) and enough adequately-pressurised (about 19 bar) water is available from the steam accumulator. The start of discharge, i.e. starting pump P2, is ordered by a signal from the ANSALDO control system when the low pressure threshold in the main steam pipe is reached. This discharge order has to be confirmed by the complete opening of the steam valve VR2 within about 40 seconds from the start of P2. Without this verification, P2 would be stopped so as to avoid pointless circulation of the hot salt. Start-up is also automatically forbidden if the temperature, in any section of the pipes, or the level of the salt in the hot tank is too low.

Discharge can be stopped voluntarily using the CAJ switch, or automatically by the above mentioned safety system.

Stop for night

To stop the storage system for a short period, such as over-night, buttons TPL-1, TPL-2 and TPL-3 are turned to the "stop" position, which changes the temperature thresholds of the regulators controlling the electrical resistances: the 430°C threshold to 275°C and the 275°C to 0°C, effectively eliminating heating.

There are also switches: MHS to put on the electrical heaters to maintain the temperature of the salt at 430°C in the hot tank and at 275°C in the cold tank; MCS to put on the electrical heaters to re-melt the salt if it has been left to freeze in the tanks.

3.6 ELECTRICAL SYSTEM

3.6.1 Introduction

The function of the electrical system is to provide power, at 20 kV, to the Sicilian grid at the local, Contrasto, hydroelectric plant. This power is generated, at 5.5 kV, by the EURELIOS turbo-alternator. When the plant is operating normally, some of the 5.5 kV supply is transformed to 380 V and supplied to the plant auxiliaries. At other times the auxiliaries bus can be fed from the grid, either through the normal transformers or directly via a 20 kV/380 V emergency transformer. In addition, limited emergency operation is possible from batteries whose state of charge is maintained by the system during normal operation.

A general outline of the electrical system appears in Fig. 3. 6. 1. The plant is normally connected to the Italian grid at Contrasto, at 20 kV, through an underground cable, a circuit breaker, A, and a 20 kV/5.5 kV transformer, T1. The 5.5 kV line is normally supplied directly from the alternator through circuit breaker B but during start-up, etc., power comes instead via breaker A and T1. The auxiliary transformer T2 reduces the 5.5 kV to the 380 volts required by the plant's other systems and made available at several specific terminal points inside the power distribution cabinets. In emergency, i. e. any failure affecting the supply via T2, the 380 V auxiliaries can be supplied from a second auxiliary transformer T3 (20 kV/380 V) provided by ENEL as a back-up source and connected to another section of the Italian grid. The T2/T3 transition is automatic as soon as the voltage fails, as well as the reverse transition T3/T2 when it comes back. The emergency electrics also includes a battery charger, battery system and inverter to power priority (and DC) users.

In short, the normal operating procedure is first to start the auxiliaries using power from the grid through breaker A and transformers T1 and T2; then to operate the heliostat field, the feedwater system, receiver, turbine and alternator; and finally to perform automatic synchronisation and closure of breaker B, thus connecting the plant to the grid. The plant auxiliaries then consume a small fraction of the generated power and the balance is supplied via breaker A and T1 back to the 20 kV grid. If through any failure breaker A opens and the plant remains operating and so supplying only its own auxiliary power, automatic resynchronisation might also be performed in a slightly different way, synchronising and closing breaker A without opening B. This makes it possible to connect the plant again to the grid as soon as the cause of failure has been eliminated.

All the electrical equipment is rather classical. CETHEL was responsible for:

- the 1500 kVA alternator (procured from Ansaldo) with the automatic synchronisation and voltage control equipments;
- the main and auxiliary transformers T1 and T2, respectively 1250 and 250 kVA, with protection devices;
- the 5.5 kV cabinets, containing the medium voltage equipment;
- the low voltage 380 V power distribution cabinets with terminal points corresponding to the main following uses:
  steam cycle and control system
  MBB heliostat field (power and control)
  CETHEL heliostat field (power and control)
  CETHEL salt storage subsystem (power and control)
  utilities such as light, air conditioning ...
- the T2/T3 automatic transition contactor;
- electrical equipment and wiring system of the CETHEL heliostats and

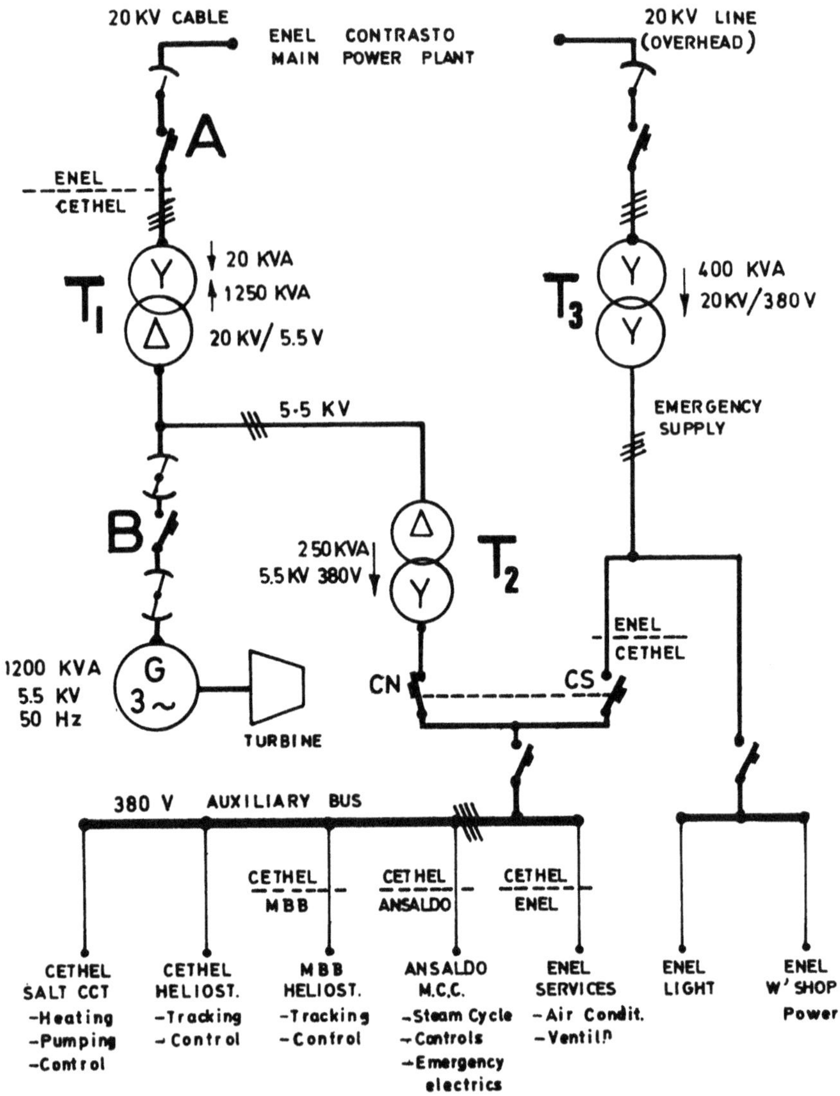

Figure 3.6.1 - ELECTRICAL SYSTEM OUTLINE

the salt storage, involving many heating resistances and protection devices.

ENEL took care of the connection to grid, providing breaker A and transformer T3.

Each Consortium member starts its own electrical installation from the CETHEL power distribution cabinets. ANSALDO in particular had many auxiliaries (pump, compressors, control systems, etc.) to be supplied and provided a specific back-up DC-current source with inverter, ensuring safe operation of the overall control system, and particularly of the protection devices and displays in the control room.

3.6.2 CETHEL electrical system

This section describes (with reference to Fig. 3.6.1) the electrical system primarily by describing its operation in some detail.

I) Preliminary conditions for operation

a) The plant is powered either by the (normal) closure of breaker A and use of transformers T1 and T2 or by the operation of the normal/emergency contactor (CN-CS) and use of transformer T3. The CN-CS contactor precludes any possibility of connecting T2 and T3 in parallel. The emergency supply allows the plant to be operated at any time except that it does not normally allow production of energy.
b) The auxiliaries necessary for the production of steam are started up.
c) The turbine is prepared for operation.
d) The turbine is started up.
e) Energy production.

II) Coupling of the turbo-alternator to the grid

Once the turbine speed has reached an 80% threshold the alternator (Table 3.6.1) can be pre-excited. When the parameters (speed, voltage, phase angle) are correct, it is possible to couple to the grid, either manually (using the ICB push-button in the control room) or, more normally, automatically.

a) Manual coupling
- the selection switch is put to the "manual" position;
- the synchronisation key is inserted and turned to "CETHEL";
- the synchronisation rack at this point becomes live and indicates
  - i) on the double-pointer frequency meter, the frequencies of the grid and of the alternator, respectively,
  - ii) on the differential voltmeter, the discrepancy between the voltage of the alternator and that of the grid,
  - iii) on the synchronisation display, the angular phase difference between the alternator and the grid.

  Coupling may be made when the frequencies are equal, the voltage difference is less than 5% and the synchronisation display is passing through zero phase difference (preferably advancing). These conditions can be achieved by pressing the "+ speed" and "- speed" push-buttons to equalize frequency and to synchronise in phase. Circuit

• Maker	ANSALDO
• Type	GSN 500 Y4
• Number of Phases	3
• Service Rating	continuous
• Power	1500kVA at 0.8 phase angle cosine.
• Voltage	5500V {+15%} {-10%} at 40°C
• Frequency	50 Hz
• Speed	1500 rpm
• Number of poles	4
• Efficiency (@ cosy = 0.8)	94.9% at full load 94.8% at 3/4 load 94.2% at ½ load
• Insulation	class F
• Overheating	class F - provision for 10% overload for 1 hour every 6 hours
• Protection	IP 23
• Bearings	Self lubricating bushes
• Total weight	5150 kg
• Exciter	type EDR 350/80, 0.31kW, 50V brushless, rotating diodes

Table 3.6.1 - Alternator Characteristics

breaker B is then closed by pressing the corresponding push-button.

b) Automatic coupling
  . the selection switch is put to the "automatic" position;
  . the synchronisation key is inserted and turned to "CETHEL";
  . the voltage, frequency and phase parameters of the alternator and grid are in this case compared by the synchrocoupler which generates the "+ speed" and "- speed" commands for the turbine's speed controller until the parameters are equal and the "close of breaker B" order is automatically issued.
  . the synchronisation equipment functions whenever coupling is being attempted and must be isolated during normal running by extraction of the synchronisation key. This is possible only when the key has been turned to "STOP" from the "ENEL" or "CETHEL" position.

III) Coupling of the plant to the grid

When the plant is in normal energy production and breaker A opens because of a fault in the grid, the turbine and transformers T1 and T2 remain in service but a command is sent to partially open the turbine bypass valve to maintain the turbine speed despite the reduction in load. After elimination of the fault coupling can again be achieved, either manually or automatically. This coupling, either manual or automatic, is carried out in exactly the same way as above except that the synchronisation key is turned to the "ENEL" position rather than "CETHEL" so that the final command is issued to close breaker A instead of B.

If breaker A opens because of a fault in the plant, breaker B will also open, the low tension (380 V) supply to the auxiliaries automatically switches (by the CN-CS switchgear) to the emergency supply (via T3) and a command is again sent to partially open the turbine bypass valve. After elimination of the fault the electrical supplies to the plant must be switched back to be through breaker A (using the CN-CS switchgear) after which the plant can be re-coupled as described in II above.

IV) Regulation of turbo-alternator

a) Speed regulator
  A bi-directional servo-motor drives the control valve, for admission of steam to the turbine in response to "+ speed" or "- speed" orders received;

b) Voltage regulator
  This regulator has three roles:
  . voltage regulation when the turbo-alternator is isolated from the grid (breaker A open),
  . cos ϕ (Phase) regulation when the turbo-alternator is connected to the grid; in this case voltage regulation is eliminated,
  . voltage equalisation operates only before coupling to the grid so as eventually to make the turbo-alternator voltage equal to the grid voltage.

V) Protection of the 5.5 kV grid

All the safety devices issue an appropriate optical and audible alarm in the control room as well as the other reactions mentioned.

a) Alternator

. mass stator (MS). Verifies stator insulation and causes, in the event of a fault, opening of circuit breaker B, ending of rotation (?) and removal of excitation from the alternator;

. power return (RP). Protects against possible feedback of power into the alternator by causing, in response to feedback of about 1% for 5 seconds, opening of breaker B;

. maximum voltage (MU). In response to a voltage increase of about 10% for 3 seconds this causes opening of breaker B, ending of rotation (?) and removal of excitation from the alternator;

. minimum voltage (mU). In response to a voltage drop of about 20% for 3 seconds this causes opening of breaker B;

. maximum current, first and second thresholds (MITA 1 and 2). In response to an increase of about 1.5 In = (normal operating current) for 3 seconds (MITA 1) or 4.2 In for 0.4 seconds (MITA 2) these devices cause opening of breaker B.

b) Protection of the T1 and T2 transformers (see Table 3. 6. 2).

	T1	T2	T3
Supplier	CETHEL	CETHEL	ENEL
Voltages	20kV/5.5kV dual use	5.5kV/400V step-down	20kV/400V step-down
Power	1250kVA as step-up 250kVA as step-down	250kVA	400kVa
Dielectric	Oil	Oil	
Coupling	star/delta	delta/star (neutral out)	star/star
Voltage Adjustment	± 5% (Adjustable off load)	± 5% (Adjustable off load)	

Table 3.6.2. - Transformer Characteristics

These transformers are protected by a pair of systems (DGTP2-type) which detect faults and actuate alarms and/or breakers.

. Temperature. This is monitored by dual-threshold thermostats. The first threshold triggers the usual alarm, at 100°C. The second, at 110°C, results in opening of both breakers, A and B, as well as the usual alarms;

. Release of gas. Gas release monitors cause only the issuing of the usual alarms;

. Overpressure. Two detectors respond to overpressure in the transformers by causing opening of both breakers A and B;

. Overload. Detectors respond to a 110% overload in nominal power of T1 by opening breaker B. In the case of T2 the protection responds to an overload of about 4 In in 0.2 seconds by opening both breakers A and B.

VI) Low voltage distribution

As mentioned above, low voltage (380 V) supply is assured by the use of two independent, alternative, supplies: via transformers T2 to T3, respectively. Normally the auxiliaries are supplied by T2, by the general safety equipment and by the CN contact of automatic transition equipment, CN-CS.

In the event of a fault in T2 the loss of voltage above contactor CN is detected and causes opening of CN and closure of CS. The short (less than 0.5 second) duration of the loss of voltage increases the security of the affected auxiliaries, particularly the 2 ANSALDO pumps supplying the receiver and the equipment for manoeuvring the heliostat field (especially MBB's which has no integral back-up power supply).

This 380 V three phase bus supplies the following users via appropriately calibrated circuit breakers:

. ENEL	60 kW	breaker	DDE,	3 x 250 A
. ANSALDO	75 kW	breaker	DDA,	3 x 250 A
. MBB	45 kW	breaker	DDMBB,	3 x 100 A
. CETHEL	30 kW	breaker	DDC,	3 x 100 A
. storage	50 kW	breaker	DST,	3 x 250 A
. regulation	20 kW	breaker	DC1,	3 x 100 A

The monitoring of the insulation of the 380 V system utilises a Merlin Gerin type TR2 controller which triggers the usual optical and audible alarms in the control room in case of a fault.

Two 380 V/220 V transformers are used to step down the 380 V supply for the regulation equipment and for the electrical tracing used to heat the salt storage pipework, respectively.

3.6.3 ANSALDO electrical system

Although each Consortium member was responsible for its own electrical conditioning, based on the 380 V three-phase supply, and these are described in the appropriate sections, Ansaldo had specific emergency responsibilities which make it appropriate to consider its electrical equipment as part of the overall electrical system.

The Ansaldo electrical equipment can be subdivided into three groups providing power, respectively, for the following purposes:

a) Main control centre (MCC), i.e. for distribution of power to equipment associated with the steam cycle, including the receiver. The MCC itself is described in Section 3.7.2. The distribution is powered from the normal 380 V, 3 phase, 50 Hz CETHEL supply, the functions of the individually switched and protected lines being as indicated in Tables 3.6.3 and Fig. 3.6.3.
b) Service board for ventilation and air conditioning. This board is powered from the normal 380 - 50 Hz supply and comprises 10 independently switched and protected lines as listed in Table 3.6.4.
c) Service board for power and light. This board provides lighting and general-purpose electrical services other than those required to operate the plant. It is fed directly from the T3 emergency supply and so

Line No.	Protection Rating (kVA)	Function
1	-	380v, 3 phase, 50Hz supply from CETHEL
2	0.25	Turbine starter motor
3	4.0	Demineralised water pump (A)
4	4.0	" " " (B)
5	5.0	Reserve
6	5.0	Reserve
7	6.0	Closed cycle cooling pump (A)
8	6.0	" " " " (B)
9	6.0	Nitrogen compressor for bottle recharge
10	6.5	Turbine lube. oil aux. pump
11	15.0	Condenser vacuum pump (A)
12	15.0	" " " (B)
13	15.0	Service air compressor (A)
14	15.0	" " " (B)
15	37.0	Receiver feedwater pump (A)
16	37.0	" " " (B)
17	3.5	Condensate extraction (booster) pump (A)
18	3.5	" " " " (B)
19	1.0	Vacuum break el./valve ? ?
20	15.0	Turbine lube. oil cleaner
21	10.0	Battery charger (see Figure 3.6-2)
22	1.0	Compressed air drier
23	65.0	Receiver and tower auxiliaries
24	1.0	Demineralised water control board
25	5.0	Inverter by-pass transformer (see Figure 3.6-2

Table 3.6.3 - MCC Power Distribution

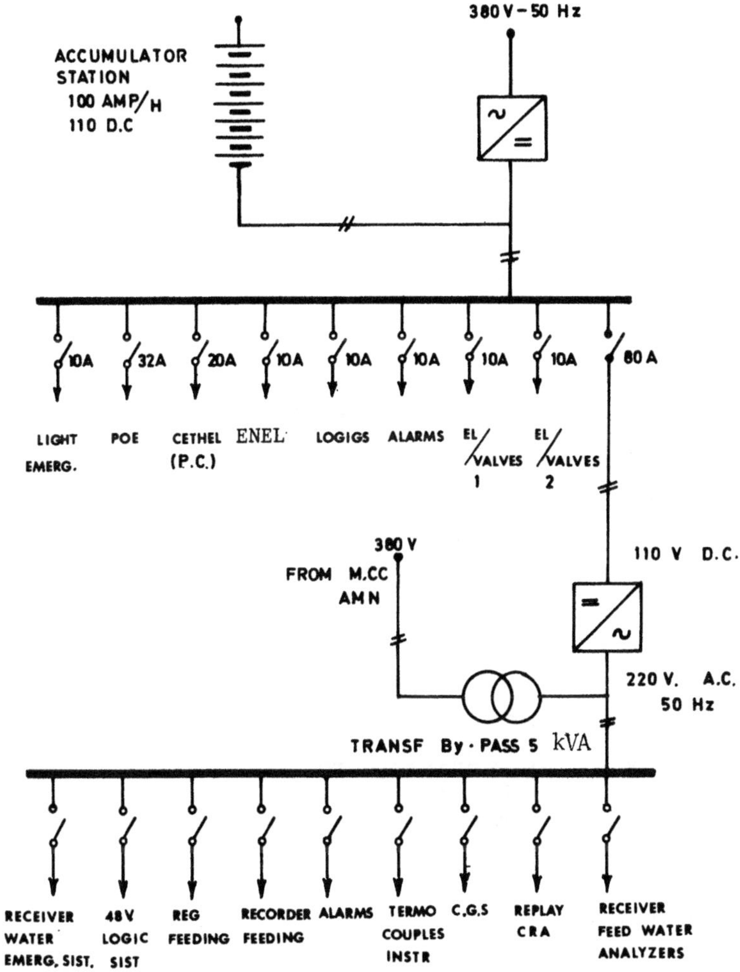

Figure 3. 6. 2. - Control Room Panel Supply

VENTILATION AND AIR CONDITIONING		
Line No.	Protection Rating (kVA)	Function
1	-	380 V, 3 phase, 50 Hz supply from CETHEL
2	0.5	Air extraction from battery room (A)
3	0.5	" " " " " (B)
4	3.0	" " " MCC " (A)
5	3.0	" " " " " (B)
6	3.5	" " " machine " (A)
7	3.5	" " " " " (B)
8	3.5	" " " " " (C)
9	50.0	Air conditioning
10	4.0	Reserve

POWER AND LIGHT		
Line No.	Protection Rating (kVA)	Function
1	-	380 V, 3 phase, 50 Hz supply from CETHEL (metered, direct from T3)
GENERAL ELECTRICAL SERVICES		
2	100	Mechanical workshop
3	70	Machine room power sockets
4	15	" " overhead crane
5	8	Service building
6	8	Reserve
7	8	Reserve
8	-	Future
9	-	"
LIGHTING (metered)		
1	5	Battery charger
2	24	Main building
3	10	Workshop and store-room
4	8	External zone
5	24	Reserve
6	-	Future

Table 3.6.4 - Electrical Service Boards

is not normally taken into account in the calculation of the energy efficiency of the plant. It comprises 15 independently switched and protected lines, metered as two groups, as listed in the second part of Table 3.6.4.

3.7 CONTROL AND COMPUTER ROOMS

3.7.1 Heliostat central controls

As indicated earlier there are differences between the control concepts of the two parts of the heliostat field. In summary these are:

. although both central units calculate the required angular positions for each heliostat, only in the case of CETHEL are these sent to the heliostats as commands. The individual MBB heliostats operate more autonomously, being given simple commands (e.g. track the receiver) and making their own calculations of the required angular positions. The MBB central control can periodically verify that the actual positions agree with its own calculations;

. the operator controls the CETHEL field via a teletype (TTY) machine and (for group or whole-field operations) push-buttons on the console. The MBB field is primarily controlled via a VDU and keyboard (normally using its TTY only for printing records of operations and faults). Of course these controls are in addition to the (identical) interfaces between the heliostat central units and the plant's general control system, GCS, used for normal plant operation;

. CETHEL heliostats can turn automatically away from the receiver, under battery power, in emergency. MBB's require power from the central unit and so do not move if such power is not available.

3.7.1.1 CETHEL heliostat central controls

Summary of CETHEL heliostat field operation

The CETHEL field is divided into four groups (Fig. 2.3) each controlled by a group controller (CDG). Automatic command and control of the field is managed by the field controller (CDC - Controleur de champ) which receives instructions from the CETHEL desk, the TTY or the general control system (GCS). As described below, each group can be commanded to any one of six operational positions; the group can be commanded in three different modes and, by using the TTY, a qualified operator can carry out a number of special manoeuvres, over-riding automatic operation and allowing maintenance, alignment tests, extraction and re-insertion of heliostats of a group, correcting time, etc.

Six operational positions

Stowage: The heliostats are fixed in a position in which they will best survive bad weather. Although initialised with the mirrors approximately face down (azimuth 0°, -84° elevation) the position is replaced by that indicated by the thumb controls on the desk as soon as transition to this mode is commanded in either of the "group" command modes. The operator should carefully select a position compatible with local weather conditions.

Safety: The heliostats are in a vertical position (elevation 0°) so that the reflected rays are directed towards the ground and no spot is formed. This state is initialised with zero azimuth but this value is replaced by that indicated by the thumb controls on the desk as soon as transition to "stowage" position is commanded in either of the group command modes.

Start: The heliostats are focussed on a point at the base of the safety trajectory, i.e. 20 m east of the base of the tower. Motion from "safety" to "start" is implemented first in azimuth and then in elevation, and vice versa.

Standby: The heliostat beams are focussed on a point at the top of the safety trajectory, i.e. 20 m east of the receiver. Motion between "start" and "standby" is constrained such that the beams from the heliostats all pass inside a 20 m diameter vertical cylinder whose axis passes through the "start" and "standby" positions.

Receiver: The spot formed by the heliostats is in the receiver.

Maintenance: This supplementary mode allows TTY (operator) control of groups and individual heliostats. The commands available to the trained operator are summarised in Table 3.7.1. If, whilst any heliostat(s) is in "maintenance" an emergency occurs which makes it necessary to move to the "stowage" position, the manoeuvre must be carried out locally, putting the heliostat(s) in manual mode by means of the switches in the heliostat electronics. In the case of strong winds, it is necessary to allow for the time required for this operation.

Three modes of command

Manual, group: Each group is commanded directly by "stowage", "safety", "start", "standby" and "receiver" push-buttons on the left side of the control desk (Fig. 3.7.1) and accepts no field commands from the right-hand side of the desk nor, except the E1 and E2 emergency alarms (E1 only when in "receiver"), from the GCS. The status of each group can be seen from the group signal lights to the left of the desk and operator and GCS messages are both available via the printer.

Manual, field: In this mode, movement of the groups is coordinated by the field controller (CDC) and is initiated by the "stowage", "standby" and "receiver" buttons on the right of the desk. The group signal lights, the printing of operator and GCS messages and the reactions to E1 and E2 alarms is as in the "manual, group" mode. The coordinates of the stowage position indicated by the thumb-wheels on the console are memorised when the manual, field button "stowage" is pressed to pass from the "standby" to "stowage" positions.

Auto, field: In this case the field is controlled by the GCS: except that the operator can intervene, using the TTY, by putting any group to "maintenance". Unless the operator has put a group into manual mode, using the TTY, the push-buttons on the console are ignored. The group signal lights operate as in the manual modes and the stowage-position thumb-wheel values are memorised when a "standby"-to-"stowage" command is received from the GCS. Transfer to a more automatic mode

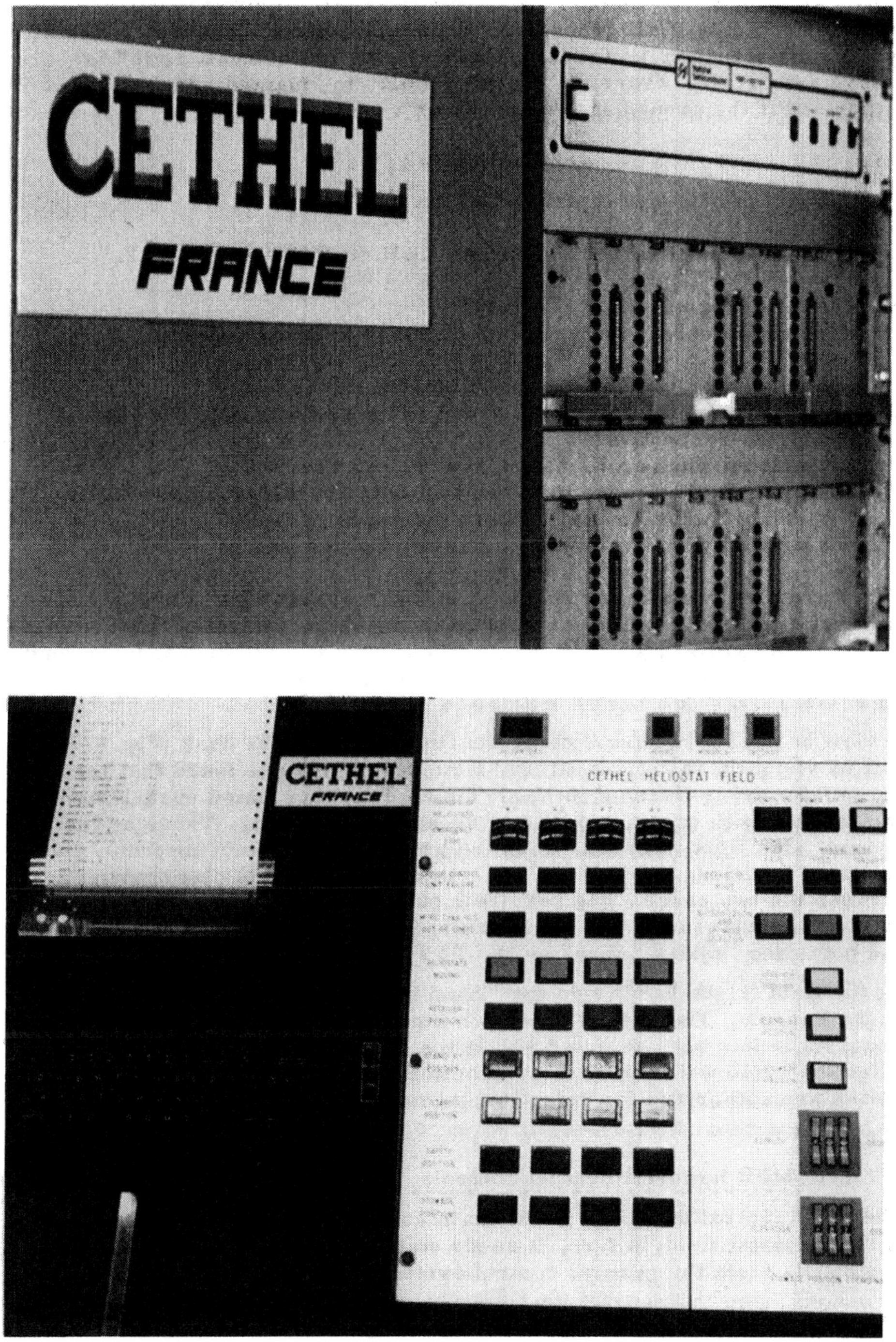

Figure 3.7.1 - CETHEL Control Desk

("manual, group" to "manual, field" or "manual, field" to "auto, field") is authorised only if all groups are in the same state ("stowage", "stand-by" or "maintenance"). Transfer "auto, field" to "manual, field" is always authorised. Transfer "manual, field" to "manual, group" is not authorised if the groups are in "receiver".

Logical status and other messages

A group status message is printed on the TTY in the form

"GROUP N, LOGICAL STATE" = .G.H.%.R.A (or M) .C.T."

with the following meaning:

T	is printed only if the group clock does not agree with that of the CDC
C	is printed only if the group arithmetic unit is faulty
A	(or M) is printed when the group is under Automatic (or Manual) control, respectively
R	is printed when an E1 alarm has tripped the group
%	is printed when more than two heliostats in the group are faulty
H	is printed when any heliostat in the group is faulty
G	is printed when an E2 alarm has tripped the group

There are also two similar heliostat status messages and a number of diagnostic printouts which can appear on the TTY, including those already indicated in Table 3.7.1. They are summarised in Table 3.7.2.

The CETHEL control console

In view of the above descriptions the illustration of the desk (Fig. 3.7.1) will be virtually self-explanatory. However, it may be noted that the time displayed is Greenwich Mean Time. The illuminated push-buttons are marked with cross, circle and square in the figure. Those on the "group" side light in green, when the group is in, or moving from, the indicated position, the yellow "heliostats moving" light distinguishing between the two cases. The red "10% out" light is illuminated when more than 2 heliostats in the group are faulty (malfunctioning or not responding) and not under "maintenance" by the TTY "HO" command.

On the field (right-hand) side the "lamp test" button tests the other lamps on the console. The green "operation mode" lamps indicate which operational mode has been selected whilst the "authorisation and op. mode selection" yellow-illuminated push buttons indicate which alternative modes are authorised for selection, according to Fig. 3.7.2.

3.7.1.2 MBB heliostat central controls

The MBB central unit (CU) is the main control device for the MBB part of the heliostat field. In turn, it sends status information to and receives commands from the general control system, GCS.

The MBB CU provides power to the heliostat field. It is connected to the heliostats via two serial data buses and two static signal lines. It commands the heliostats via the "command data bus" and receives requested status information from a selected heliostat via the "response data bus".

ROUTINE COMMANDS	
COMMAND*	RESULT
FT	Printout of Julian day and time of the central clock in the format: "MASTER CLOCK = xxxxx JD xx H xx MN xx S"
GL,N	Printout of logical status of group N in the format: "GROUP N, LOGICAL STATUS = X.X.X.X.X.X.X.X.X."
GT,N	Printout of Julian day and time of the group N clock in the format: "CLOCK, GROUP N : xxxx JD xx H xx MN xx S"
GA,N	Transfer of group N to 'Auto'
GM,N	Transfer of group N to 'Manual'
HL,N,NN	Printout of logical status of heliostat NN of group N in the format: "GROUP N, HELIO NN, STATE = .X.X.X.X.X.X.X.X A = .X.X.X.X S = .X.X.X.X"
HP,N,NN	Printout of position of heliostat NN of group N in the format: "GROUP N, HELIO NN, POSITION = A = xxx xxx S = xxx xxx"
HR,N,NN	Resetting heliostat NN of group N

COMMANDS FOR TRAINED OPERATORS	
COMMAND*	RESULT
FS,SS,MM,HH,JJL,JJH	Reset central clock time
GI,N	Re-insertion of group N
HI,N,NN	Re-insertion of heliostat NN of group N from maintenance or alignment
GO,N,±EEE,±AAA	Transfer group N to 'maintenance', position EEE, AAA degrees
VT,N,±XXXX,±ZZZZ	Correct target vector for 'receiver' by XXXX cm East, ZZZZ cm up
V1,N,±XXXX,±ZZZZ	" " " alternative target 1
V2,N,±XXXX,±ZZZZ	" " " " " 2
VA,N,±XXXX,±ZZZZ	" " " 'standby'
VD,N,±XXXX,±ZZZZ	" " " 'start'
HO,N,NN,±EEE,±AAA	Transfer heliostat NN of group N, to 'maintenance', position EEE, AAA degrees
HE,N,NN,±XXXX,±ZZZZ	Transfer heliostat NN of group N to 'alignment' on the given vector, $\lvert X \rvert \leq 9999$cm, $\lvert Y \rvert \leq 9999$cm

*Each command must be terminated by 'carriage return'.

Table 3.7.1 - CETHEL TTY Commands

1. MASTER CLOCK (response to command FT).

2. GROUP STATUS (response to command GL or to any change of group status).

.G	.H	.%	.R	.A (or M)	.C	.T
Emergency E2	Heliostat fault	More than 2 H'stat faults	Emergency E1	A-Auto M-Manual	Calcu- lator error	Clock error

3. CLOCK, GROUP N (response to command GT).

4. OPERATOR ERROR (response to operator error in TTY input).

5. GROUP N COMMAND ERROR (response to communication errror).

6. GROUP N OFF (results from lack of response to CDC commands).

7. HELIOSTAT STATUS (response to command HL or to any change of heliostat status).

HELIOSTAT STATUS 1

.A	.A	.R	.S	.B	.M
Emergency E2	Thermal Alarm	Moving	Secteur?? fault	Battery fault	Manual mode

HELIOSTAT STATUS 2

Azimuth				Elevation			
.O	.R	.H	.F	.O	.R	.H	.F
out of action	speed error	fast move- ment	position error	out of action	speed error	fast move- ment	position error

8. HELIOSTAT POSITION (response to command HP).

9. INCORRECT MASTER CLOCK, TIME SETTING (response, at CDC switch-on, if master clock battery has discharged).

10. FIELD CONTROLLER ON (response to CDC switch-on).

Table 3.7.2 - CETHEL TTY Messages

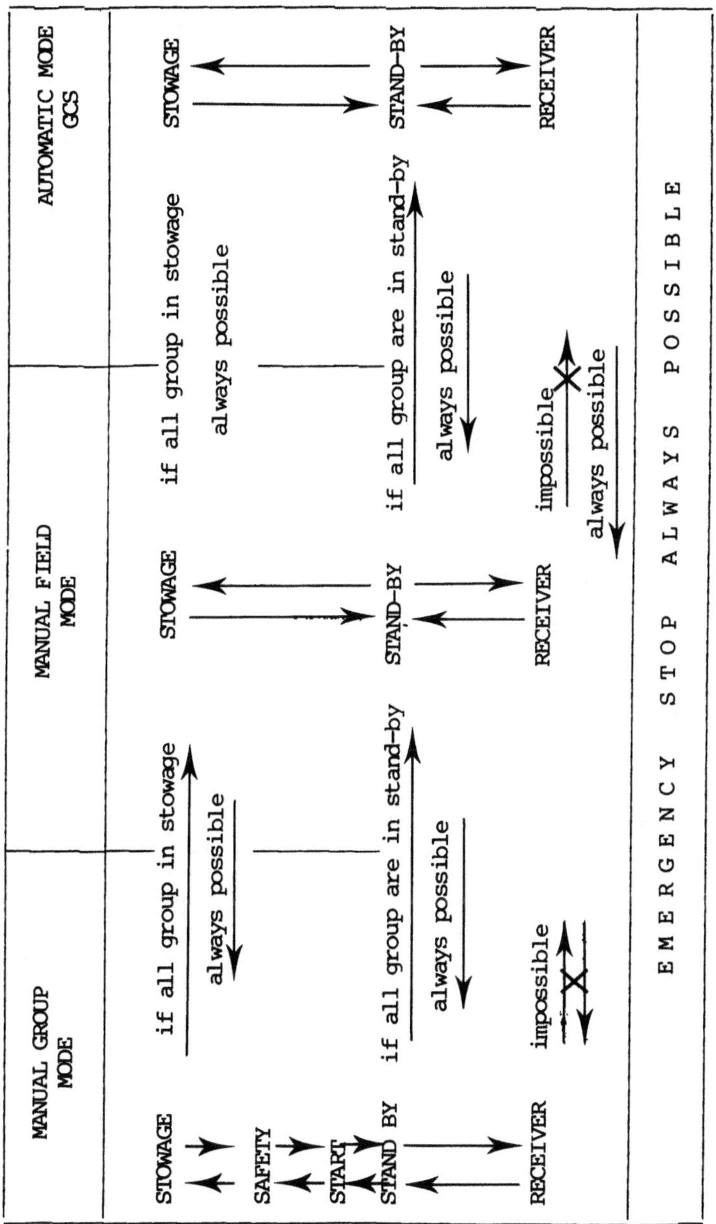

Figure 3.7.2 - CETHEL Mode Transitions

For each heliostat group two "emergency signals" are provided which control the movement of the heliostats in the azimuth (E1) and the elevation axis (E2) directly in case of a serious failure even if this occurs in the serial bus system.

Dependent upon automatic mode commands from the GCS or on manual commands initiated by the operator, the MBB CU sends corresponding commands to a single heliostat, to a group of heliostats or to all heliostats (date, time, position, status requests). The CU also continuously verifies the correct function of each heliostat by periodically requesting status information. As an additional control the actual positions about the two heliostat axes are requested from each heliostat after the heliostat indicates that the "receiver position" has been reached. The CU calculates for the heliostat the correct values and compares these with the actual values received from the heliostat. If the comparison indicates a difference, an error message will be produced. In any case all status information obtained from the heliostat field is assembled by the CU and indicated to the operator. The heliostat field status is also provided on specified data lines to the general control system.

The MBB central unit consists of two main units, the MBB console in the control room and a power cabinet in the "sottoquadro". A block diagram and photographs appear in Figs. 3.7.3 and 3.7.4.

The MBB console houses the electronic circuitry of the CU. One active electronic unit and a second one in cold standby redundancy are installed in the base of the console. The electronic units consist almost entirely of off-the-shelf microcomputer systems. The central unit processor (CUP) is an Intel 8080 processor-based unit containing the arithmetic sub-unit (which checks the heliostat position calculations), 16 kbytes of EPROM and 2 kbytes of battery-buffered CRAM, an interrupt unit, counters and interfaces.

The console itself contains a control desk, an Intercolour 8001 CRT plus keyboard and an HP 2635A line printer. The control desk contains (one per group) illuminated switches for power on, manual override and emergency deviation and also signal lights showing the status of each group (10% out of limits; moving; in receiver position; in standby position; in stowage position). It also has switches controlling power to the processor, to the interface sub-unit and to the whole field, initialisation and emergency reset and also signal lights showing the existence of GCS command sequences and of errors in the field. The CRT is normally used to display information to the operator for the dialogue between him and the central unit processor. Instead of the CRT and keyboard the line printer also may be used for this dialogue. It also provides a permanent record of error messages and other service information.

Other sub-units of the MBB CU are:

. Timing unit, a quartz-controlled adjustable oscillator and associated counters, supplied by an accumulator-buffered supply (in the AKKU, see below), to provide stability for about 7 days. It can be synchronised periodically to the CETHEL clock.
. Emergency unit. Able to operate independently of the central unit

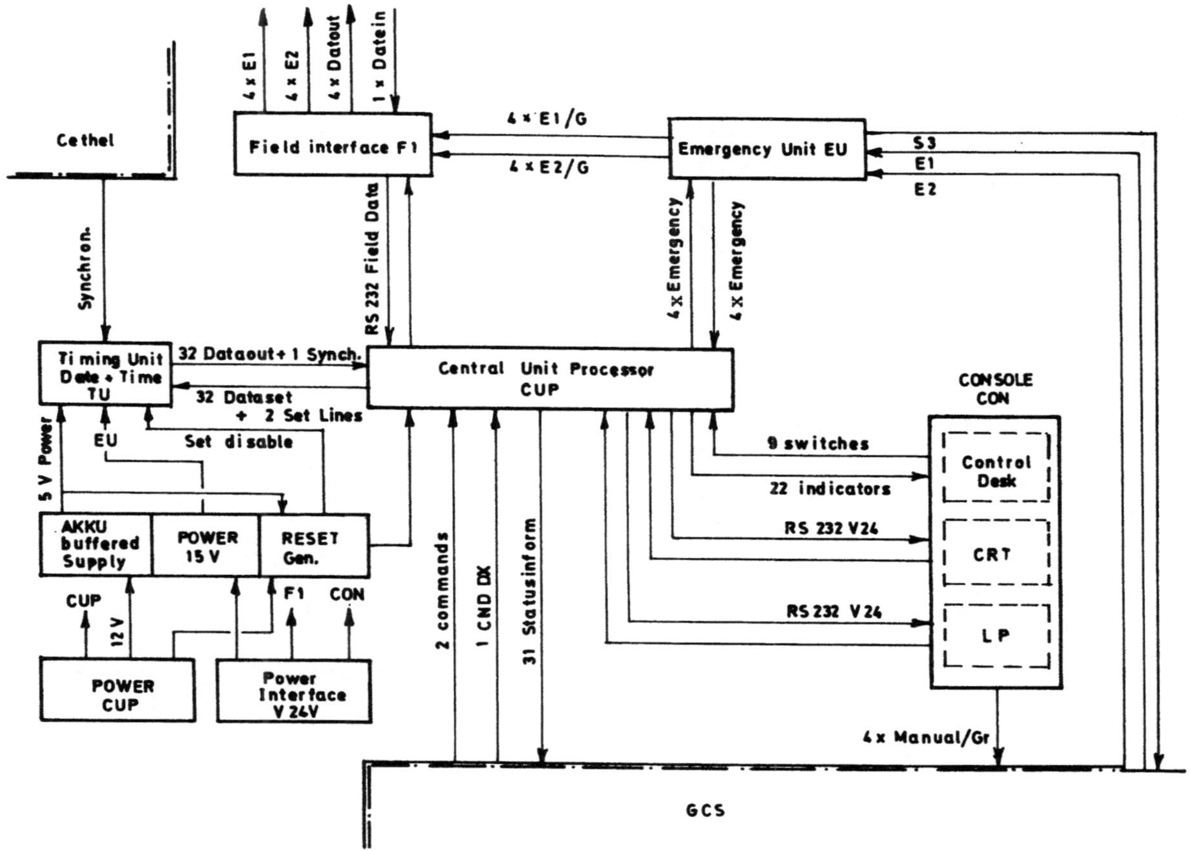

Figure 3.7.3 – BLOCK DIAGRAM OF MBB CENTRAL UNIT

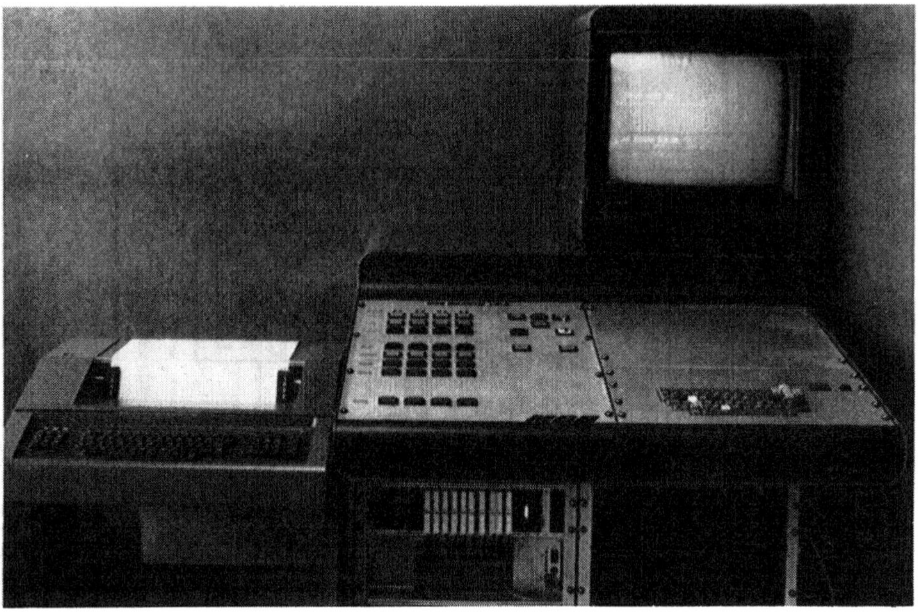

MBB Central Unit Console
(Control Room)

Power Cabinet for MBB Heliostats
(in "Sottoquadro")

Figure 3.7.4 - MBB Central Unit

processor, this unit handles the automatic emergency sequences required in response to E1/E2 commands from the GCS or to an emergency command from the CUP.

- Field interface unit which buffers all signals to and from the heliostat field and includes appropriate short circuit and over-voltage protections.
- AKKU. This contains an accumulator-buffered 5 V supply for the timing unit and for reset generation; a capacity-buffered 15 V supply for the emergency unit which avoids an "emergency" sequence if there is a short (about 1 second) power loss; a reset pulse generator to ensure correct status for the CU and the timing unit after power-up.

Detailed functions, operations and interfaces of the MBB central unit are specified in the "Operators Manual" but they may be summarised as follows.

In the absence of commands, the central unit continuously scans the entire MBB field, group-by-group, heliostat-by-heliostat, verifying that there are no discrepancies between expected values and those measured. If a discrepancy is detected the CU requests a "failure word message" from the affected heliostat. This results in the printing of a message on the line printer identifying it as a data handling failure; arithmetic logic unit failure; lack of power to drive servos; elevation or azimuth end positions reached or discharge of the small battery intended to preserve the contents of the heliostat control logics RAM. Alternatively, when a group is tracking the receiver, the position of the heliostat is printed out if it differs from that calculated by the CU processor's arithmetic unit.

Summary tables can also be displayed, in response to "0" commands (Table 3.7.3), which show the status of every heliostat in terms of whether or not it is under control; tracking the receiver; disabled from accepting group commands (to allow individual heliostat maintenance for example) or "cancelled" (i.e. automatically inhibited from responding on the response bus because of a short circuit which would otherwise block the entire bus).

If a "no bus answer" error is detected for a whole group, an E2 emergency message is issued (taking the group towards stowage position).

The "0" command has already been mentioned. As well as responding to group commands from the GCS, the MBB field can be commanded from the console, using the commands shown in Table 3.7.3. The GCS group commands include "sequence" commands as a result of which a group command is sent out to each group in turn over a period of time. Similarly the GCS emergency commands E1 and E2 activate the emergency lines regardless of the status of the CU processor.

Interface between the GCS and the MBB central unit

Via coded parallel data lines the CU receives commands from the GCS to move the heliostats into certain positions (stowage, standby, receiver) and sends heliostat field status information back to the GCS. The output and input circuits of the CU provide complete insulation by opto-couplers and, particularly, relays.

Command	Function
R (Adr)	Go to Receiver
T (Adr)	Go to Standby
S (Adr)	Go to Stowage
G(Adr)A(Grad) (Min)	Go to defined azimuth
G(Adr)E(Grad) (min)	Go to defined elevation
(moving starts after GE command)	
M (Adr) (xy)	Move with defined speed and direction
x/y	
0 Stop Azimuth/Elev'n	
1 Slow West/up	
2 Slow East/down	
3 Fast West/up	
4 Fast East/down	
A (Adrs)	Go to alignment
B (Adrs)	Send alignment table
P (Adrs)	Request actual position
Q (Adrs)	Request calculated position

Command	Function
W (Adrs)	Request status word
F (Adrs)	Request failure word
Y	Display date, time
O 1	Control-parameter table
O 2	Tracking-parameter table
O 3	Group-cmd-disable-param.table
O 4	Cancel-parameter table
E	Output status table
H	Output alignment table
K(Adrs)(Value 1,2 .. n)	Set alignment table
L (Adrs)	Enable group address
N (Adrs)	Disable group address
C (Adrs)	Cancel heliostat
J (Adrs)	Reset cancel command
V (hour,min)	Set time
D (day,month, year)	Set date

Note: (Adr) stands for either a single heliostat address, 1-112, or a group address 251, 252, 253 or 254 (operator's option). (Adrs) stands for a single heliostat address, 1-112.

Table 3.7.3 - MBB Central Unit Console Commands

The following commands can be initiated from the GCS via the MBB CU:

. a specific single group into stowage position, standby position or to illuminate the receiver;
. complete field to stowage, standby or receiver position in defined sequences, group by group.

Sequences to and from the stowage positions are introduced to avoid high peak power demands. The sequences for movement between standby and receiver position are related to the allowed temperature gradient of the receiver.

Two additional emergency commands GCS-E1 and GCS-E2 are generated by the GCS. The CU activates the corresponding field emergency lines independent of the internal CU status or any switch status on the console desk. After receiving the GCS-E1 signal a time-limited emergency signal E1 for the whole field will be generated by the CU causing a high-speed eastward movement about the azimuth axis. The TCS-E2 signal initiates in the CU to generate first the E1 field signal for a limited time as above and then the generation of a steady E2 field signal which lowers the mirrors by rotation about the elevation axis.

MBB power interface

The heliostat field is powered from the 380 V three phase AC-grid. The power cabinet of the CU converts this through a 45 kW, 380/380 V star/star transformer (which assures complete isolation from the CETHEL electrical supply) to provide the 220 V supply on which the CU operates and the 380 V power supply for the heliostat field. The power cabinet provides all necessary switching, fusing and distribution devices. The following values are specified as power demand for the heliostat field:
Input power:

. voltage	3 phase 380 V $\pm$ 10%
. power consumption:	
whole field during tracking, average	4 kW
whole field maximum (e.g. emergency)	35 kW
central unit max.	0.5 kW

Apart from small batteries to avoid reset or loss of data in the event of a brief interruption of power neither the heliostats in the field nor the CU include any significant power storage capability because of the inspection and service requirements, limited life time and need for additional charging circuits. The plant, therefore, normally provides emergency power for the periods when a power drop or switch-off occurs on the normal grid and can also tolerate the MBB field's remaining stationary until the sun has moved enough for the image to leave the receiver.

Interface MBB central unit/MBB heliostat field

The interface between the CU and the heliostat field consists of cables for power and for data. The data connections have two separate lines for the "emergency signals" and the two serial bus lines (command bus and response bus) using 110 bits/sec RS 232, CCITT V.24 (DIN 62 020) standard signals.

3.7.2 Master Control Centre (MCC)

General

The MCC provides control and supervision of the whole plant including receiver, steam cycle, heliostat field and storage. For the definition of its present configuration ENEL referred whenever possible to usual procedures adopted in thermal power stations.

3.7.2.1 Information and control layout

A) Foreword

Before starting to explain in detail the control and supervision modules it is important to understand the basic interconnections and relationship among them, shown in Fig. 3.7.6. In this figure boxes represent plant components or electronic equipments and lines mean flow of analog or digital signals. It is also appropriate to point out that practically all the relevant signals shown will (by the end of Phase D) be routinely collected and processed in real time by the supervision computer (called "DASS").

B) System outline

The SRES module (solar radiation elaboration system) receives signals, F1...F20, from 20 fluximeters welded on the boiler pipes and distributed (8/8/4) on 3 different layers, orthogonal to incoming radiation flux.
It also receives signals, P1...P4, from four sun tracking pyroheliometers (i.e. not incident radiation transducers). The SRES processes those signals and outputs into an analog signal, called Qer, the best available estimate of actual heat flux entering into the process fluid.
The SRES also drives a complete control panel (in vertical panel P611 in the control room) which allows the operator to visualize the most important variables and to eliminate faulty sensors.
The output signal Qer is monitored on the control desk and fed into the regulation subsystem where it plays a fundamental role as driver of feed-forward control actions during sun-radiation disturbances.
The meteo measurement system module measures and collects meteorological data, viz.

. wind speed and direction at 2, 10, 50 metres altitude;
. temperature, atmospheric pressure and humidity;
. global sun radiation,

and presents these data on pen recorders and digital meters fitted into the P611 vertical panel.
The 10 and 50 m wind speed signals are also processed by a special non-linear threshold circuit which closes pre-alarm contacts to warn the operator and to trigger the E2 emergency signal which in turn deviates the whole mirror field to the stowage position.
The "Command generation for MCUS's" module (or "Code") generates the digitally coded commands to drive the two mirror fields. The Code can:

. select MBB or CETHEL or both fields, under operator command;
. start or stop the deviation sequence (from receiver to stand-by) for selected field(s), under operator or regulation system command;
. start the focussing sequence (from standby to receiver), under operator command only.

Figure 3.7.5 - Master Control Centre (MCC)

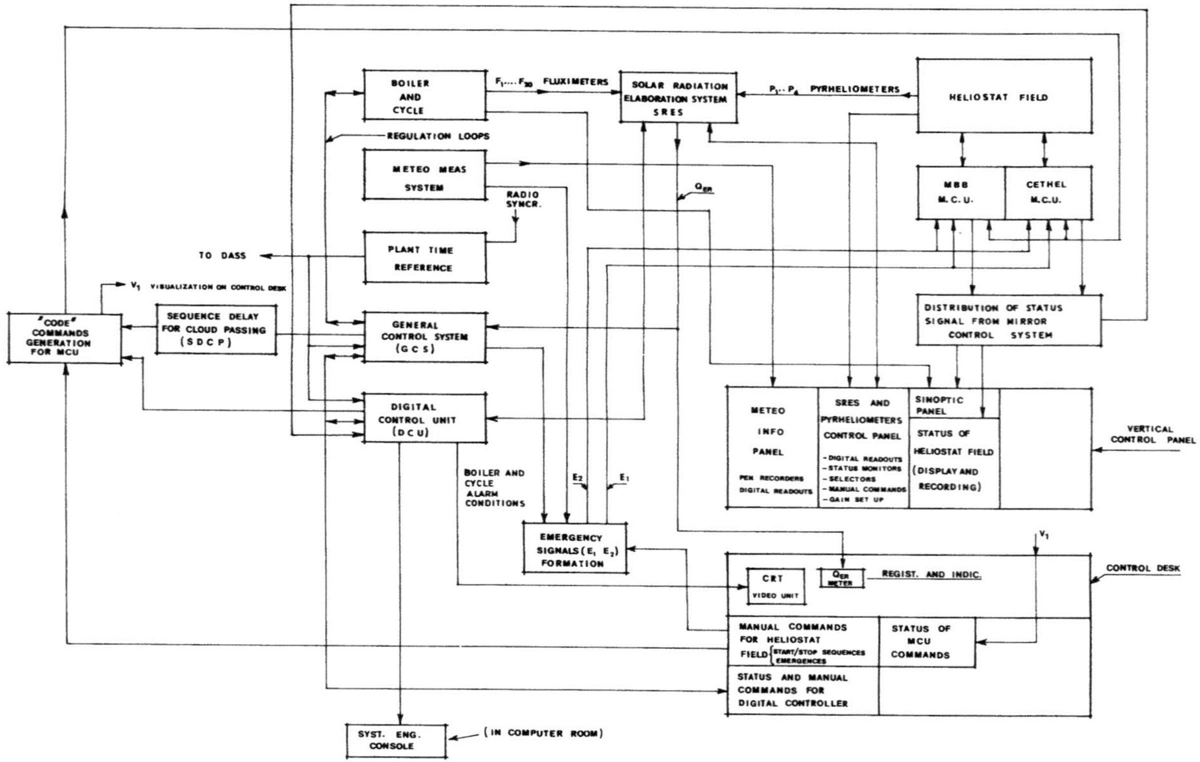

Figure 3.7.6 - MCC : Control & Supervision Block Diagram

Regulation commands come from the electronic module which detects a control system request for too low a feedwater flow rate, as happens if a cloud obscures the sun for more than 3 or 4 minutes.
If the sun returns before the deviation sequence has finished, the Code generates a "Stop sequence" command and allows the operator to restart the focussing sequence as soon as possible. The operator's commands come from push-buttons located on Control Desk.

The "distribution of MCU's Status" module receives the contact closures from the Mirror Control Units, giving the actual status of the mirror fields and (by a suitable set of relays) repeats and distributes them to the synoptic panel, for driving the lamps indicating the status of mirror groups, and to the event pen recorder for permanent storage of mirror group movements.

The General Control System (GCS) and the Digital Controller Unit (DCU) play fundamental roles as basic and more advanced control devices, respectively. Their functions are described in more detail in par. 3.7.2.2. Here it is sufficient to point out that they are connected to all other relevant systems and can be considered the "centroid" of the whole plant.
In particular, the E1 emergency signal is triggered by any of a set of boiler and steam cycle "danger" conditions.

The "Plant clock references" module:
. keeps the actual timing of the whole plant synchronised with the reference signals broadcast by the Italian network (RAI);
. drives the alfa numeric calendar and time displays located in the control and auxiliary rooms;
. distributes on a serial two-wire line (300 Baud, RS232 standard) a message containing actual time information and other relevant plant data of interest, in particular to the "DASS" system.

Vertical panels, control boards, CRTS' System Engineers Console, etc. implement the man-machine interface and allow supervision and control of the whole plant by the operator team. (Note that activation of the "DASS" system (foreseen for July 1982) will change in some way the role of control room instrumentation.) The instrumentation and the organisation of those panels is generally conventional. For some special functions like Qer formation, pyroheliometer control and monitoring, wind speed threshold, "CODE" control and monitoring and Mirror Control Unit status, special panels and monitor devices have been developed. Those devices are described in the following and in related documents.

C) Control console

This is in four sections, each housing the controls for one part of the plant:
i) Duplicates of the high level (group) heliostat controls. Once the heliostat controls are "ON", the MCC duplicates are sufficient for normal and emergency operation. The operator then needs to use the separate MBB and CETHEL control consoles only for non-routine activities.
ii) Controls for regulation of the receiver-feedwater flow rate and the attemperators/desuperheaters in the receiver.
iii) Controls for the thermal performance of the cycle and for the turbine.
iv) Controls for the auxiliaries such as the closed cycle cooling system,

compressors, vacuum pumps, and salt storage system.

The "central control" computers for the heliostat field and the control consoles used for experimental or maintenance operations with heliostats have already been described in Section 3.7.1. These consoles are alongside the MCC control console and can be reached very easily by the operator if necessary in case of non-standard (but not emergency) operation.

D) Vertical boards

The vertical boards are in six sections:

i) Meteorological signals.

ii) Controls and displays for the solar flux detectors (pyroheliometers) in the field.

iii) Controls, displays and alarms for the salt storage system (see Section 3.5.2) and for the alternator and its auxiliaries.

iv) A synoptic display of the status of the main components of the plant, together with alarm signals related to the receiver-feedwater, main steam, etc.

v) Temperature recording equipment for the receiver tubing and for the turbine, together with alarms for the turbine, the rest of the steam cycle and auxiliary machinery.

vi) Measuring, recording and alarm signals for the 200 kV and 5.5 kV power system including meters to record the energy produced by the alternator and that absorbed by the power plant's auxiliary systems.

3.7.2.2 Control concepts, overall description

The concept of the plant (extended) control system is described by the general scheme of Fig. 3.7.6. The role of the different components can be summarised as follows:

The Mirror Control Systems (MCS) supplied by the heliostat manufacturers ensure correct tracking of the mirror field and drive the heliostats along the prescribed trajectories. They supply the essential information about the field state both to the operators and to the automatic control devices. MCS commands can be issued both by the operator, through the mirror control consoles, and by the GCS (general control system), to which the MCS appears as a slave. The MCS consists of two centralised digital units (one of MBB and one of CETHEL) and of a distributed set of digital units, each unit devoted to the control of one heliostat. A more detailed description of MCS has been given earlier in this report.

The receiver and thermal cycle control system (RCS) consists of two separate parts: the receiver and thermal cycle regulation system (RRS) and the manoeuvre automation system (MAS). The RRS includes separate local regulators, of quite standard type, for the conventional components of the thermal cycle (turbine, condenser, flash-tank, thermal storage, etc.) and the integrated receiver regulation system, devoted to the control of pressure and temperatures in the receiver.

The essential functions of RRS components relevant to the receiver and steam cycle control are described in Fig. 3.7.7. In phase C, the RRS has been completely realised in analog technology. The MAS, though conceptu-

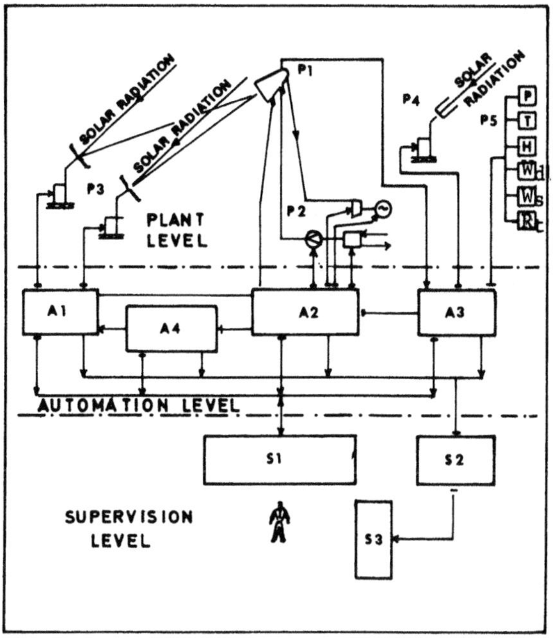

KEY

P1 Solar receiver, steam generator
P2 Thermal cycle
P3 Mirrors field
P4 Solar radiation measurements
P5 Meteo measurement devices
A1 Mirror control system
A2 Receiver and thermal cycle regulator
A3 Thermal flux and meteo measurement system
A4 Manoeuvres automation system
S1 Operator bench and vertical panels
S2 Supervision process computer
S3 Supervision consoles
W_d Wind direction
W_s Wind speed
R_t Total radiation
P Pressure
T Temperature
H Humidity

Figure 3.7.7 a

GENERAL CONTROL SYSTEM (GCS)

Structure

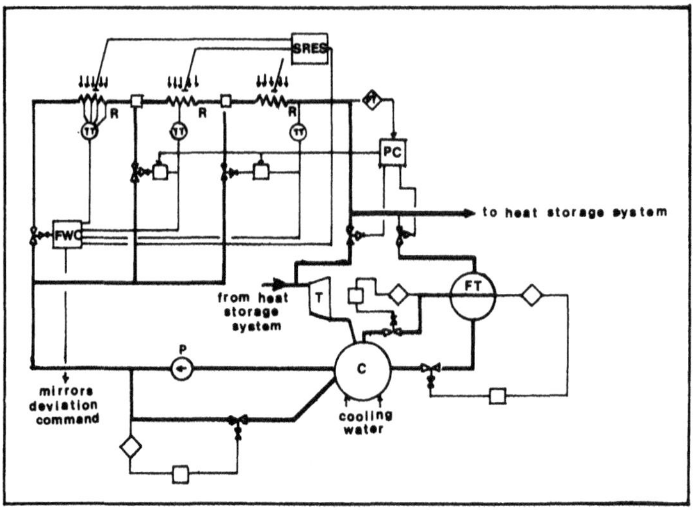

KEY

TT	Temperature Transducer
PT	Pressure Transducer
FWC	Feed Water Controller
ESC	Evaporator Spray Controller
SSC	Superheater Spray Controller
SRES	Solar Radiation Elaboration System
T	Turbine
C	Condensor
P	Feedwater Pump
R	Receiver
FT	Flash Tank

Figure 3.7.7 b

ally defined in Phase B, is to be implemented after Phase C but its interactions (as in Fig. 3. 7. 6) have been taken into account in the design of the RRS which includes the interfaces necessary for correct cooperation between the two control subsystems.

Special and peculiar features of the RCS are the interactions with the mirror field and with the thermal flux measurement system. These have been conceived to allow the RCS to command the MCS (e. g. for sequential and emergency mirror deviation) and to receive a continuously updated estimate of the total thermal power entering the receiver (from the SRES). Details on these special devices and their operation are given in sections 3. 7. 2. 3 and 3. 7. 2. 5.

A supervision process computer (SPC), and supervision consoles (SC), will be devoted to overall plant supervision, i. e. to plant data acquisition, monitoring and displaying, plant performance evaluation, operator alarming of plant malfunctions, plant data storing for post-mortem analysis, etc. Installation is planned for Phase D of the project.

The basic concepts of the RRS can be summarised as follows:

- Receiver outlet pressure is controlled, either by means of the turbine valve during normal operation or by means of the by pass valve during the plant start-up and shut-down, to avoid unwanted motion of the evaporating zone in the receiver.
- Receiver outlet temperature is controlled by modulating the feedwater mass flow rate in both receiver tubes and the attemperator spray valve positions, based on fluid temperature measurements at intermediate points in the receiver to anticipate control actions. As usual, feedwater flow is used for low-frequency regulation whereas spray flow is used for higher-frequency control.
- Receiver control in the event of large solar radiation disturbances (e. g. cloud passing) relies essentially on the feedforward control action obtained from the thermal power estimate computed by the SRES device.
- The feedwater control system (FWS) allows the temperature control system to be independent of the feedwater valve characteristics and of the pressure drop variations. In particular, to speed up the feedwater control, FWC has been designed to interact with the pump header pressure control.

An extensive and detailed description of RCS concepts and algorithms was described in March 1980 in a report by ENEL.

The RRS was tested at the manufacturer's facilities using a digital simulation of the plant. The structure of the simulation and the complete testing procedure have been documented by ENEL and the manufacturer and their essential features published.

3. 7. 2. 3 Special control features

Two aspects of the control system design are very peculiar to this application. The first is related to the need to deal with very long transport delays, which affect the receiver dynamics. This calls for the use of special feedback control actions which cannot be realised by traditional PID regulators. Therefore, the Manoeuvres Automation System (MAS), conceived in Phase B but not included in Phase C of the project, is intended to implement not only sequential automation of basic operating procedures but also some special digital control algorithms which are superimposed on the analog regulation system through suitable interfaces. In fact, the MAS is to be realised by Digital Control Unit (DCU) and the analog regulation has been designed to be compatible with the DCU outputs and inputs. The concept and performance of the digital control system and the results of simulations have also been reported in detail by ENEL.

The second special feature of the RCS is represented by its role as master control for the MCS. In this context, the RCS commands mirror sequence deviations when the thermal power to the receiver is not sufficient to keep the steam generator above the technical minimum. Since those events are normally due to cloud, which can pass more quickly than the several minutes required for depressurising the steam generator, the sequence deviation is commanded only if thermal power is lacking for a sufficiently long time.

This concept has been incorporated in the control system by means of two different devices connected in series:

. a "code device" (CD) which translates the simple logic command of the RCS to an MCS-compatible form by a suitable protocol;
. a "sequence delay device" (SDD) which transfers the actual deviation command to the MCS only if a thermal power lack is recognised by the RCS for a sufficiently long time interval (e.g. 5 min.).

The choice of the technical minimum, of the detection criterion for loss of power and of the sequence delay value have been based on receiver safety arguments and suitable dynamic simulations using the ENEL mathematical model.

3.7.2.4 Plant operation: manual -v- automatic options

The plant control status at the end of Phase C is essentially as follows:

. The receiver regulation system is kept in complete automatic operation after the start-up of the plant. Only receiver pressure regulation, pump header pressure regulation and feedwater control system are necessary and effective during the start-up. The temperature regulations can be put in automatic regulation only when the steam temperature at the receiver outlet is above about 400°C.
. The plant manoeuvres during start-up are performed manually by the operator at the control board except for a limited number of operations which should be done directly on the plant (seal steam system start-up, arming of the turbine and first opening of the turbine control valve, pre-excitation of the alternator, closing of the flash-tank regulating valves).
. Manual activation of the salt storage charging and of the steam storage filling.
. Possibility of commanding the mirror field from the main control board through the Code Device (CD). This enables the operator to activate standard sequence deviations of the mirror field without operating on the mirror control board.

A thorough description of the operating procedures, including specification of automatic and manual controls, can be found in the provisional operating manual (in English) compiled before the first plant start-up and in the updated operating manual (in Italian) which was developed from the earlier one after three months' operating experience. A summary is given in Chapter 4 of this report.

As a general remark, it can be said that the present semi-automatic operation of the plant during start-up is not a very significant drawback because the very slow dynamics of the solar receiver allow the operator adequate time for all the required manoeuvres. Full automation of the start-up manoeuvres would imply some reduction in operating personnel but little saving of time (perhaps 20 minutes).

3.7.2.5 Special control devices

The meteo measurement system consists of four pyrheliometers (P1... P4) installed on steel pedestals 5 meters high around the mirror field. Their operation and protection against bad weather by a coverage is controlled from the control room. Two wind speed measurements are taken from a mast at 10 and 50 meters height above ground. Besides the registration via pen recorders the device provides two logic signals which indicate that the adjustable threshold T1 and T2, corresponding to wind speed preset limits, have been overcome. Under automatic operation conditions (S1 on AUTO) the alarm results in heliostats going to stowage position as required for safety of equipment. Under manual conditions, the operator can over-ride under his own responsibility.

Elaboration of Qer by the Solar Radiation Elaboration System "SRES" is illustrated in Fig. 3.7.8. It calculates a sort of weighted sum of 20 individual fluximeters, taking into account the nominal flux values and giving the output signal as a percentile of the nominal value. It provides a high degree or redundancy because many fluximeters can be put out of service without affecting the value of the output signal. It calculates also a Qer estimation from pyrheliometer signals, which can be used for comparison and/or monitoring purposes, or, in case of total failure of fluximeters, as control signal for water flow rate regulation. It provides some manual and automatic monitoring possibilities which allow the system engineer to visualise the intermediate values of Qer's computation and which allow the DASS to digitally interrogate the "SRES" for having back the fluximeters values and the final computed Qer value.

Flux monitoring is achieved via a manual selector, mounted on vertical panel under the "SRES" device, which allows the operator to visualise on a 3 digit numerical display the actual value of each fluximeter output, up to a maximum of 40 independent signals. Up to now we have monitored by this device the 20 fluximeters also giving input to the "SRES" apparatus, in order to check the "SRES" input and to find wrong fluximeters. Because the above mentioned point by point manual monitoring does not give a synthetic view of a whole fluximeters set, we developed a special instrument (called "flux mapper device" (FMD)), which:

- accepts up to 48 fluximeter signals;
- converts them in logical pulses, whose duration is proportional to actual analog value;

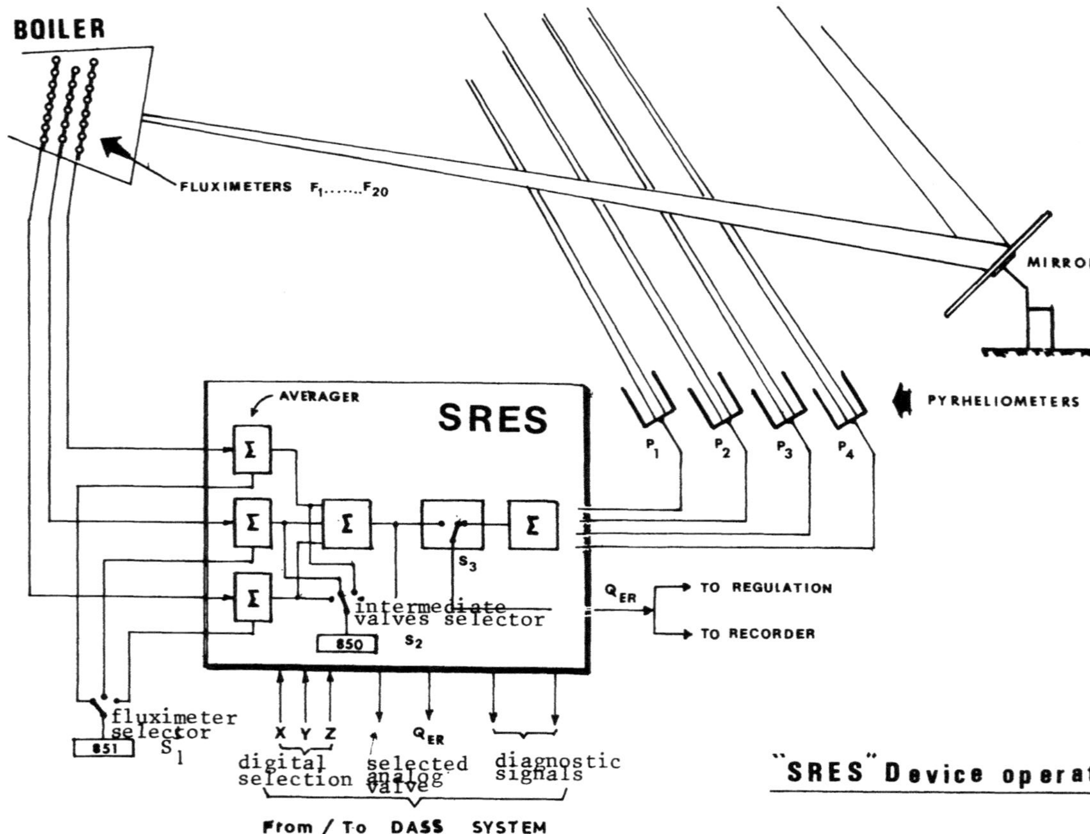

"SRES" Device operation

Figure 3.7.8

. presents them as a vertical "bar graph" diagram on a multipen event recorder;
. repeats automatically this kind of measure with a pre-defined time interval, adjustable from few minutes up to 6 hours; it drives also the paper advance and pen action of pen recorder.

The output diagram gives an immediate insight of actual flux distribution: furthermore, comparing a time series of those diagrams (for instance, during morning start-up or in case of cloud passing), people can gain a suitable knowledge for better understanding the dynamical behaviour of the receiver and related plant components.

3.8 CIVIL WORKS

3.8.1 Site selection

The site, selected by ENEL after a search covering different places in Southern Italy, is located in Contrasto, in the Commune of Adrano, Sicily, about 40 km north-west of Catania. The reasons for choosing this site are:
. good exposure
. easy access
. easy connection to the national grid
. availability of cooling water
. proximity to an existing ENEL hydroelectric station
. proximity to an international airport.

The overall area available is about 310,000 m^2 with a mean elevation of 220 m above sea level and a slight natural inclination north-east to south-west. The plant borders on regional road no. 16 (SS 575) to which it is directly connected, its position being shown in Fig. 3.8.1.

3.8.2 Survey and site preparation works

A soil investigation campaign performed in 1978 disclosed a basalt stratum beneath a subsoil and topsoil formation varying in thickness between zero and about 4 m. The subsoil consisted of sand and clay silt containing small rock fragments and was topped by a layer of humus. The land profile was quite irregular but was levelled off by removing all the topsoil (approximately 70 cm) and part of the subsoil. Final grading of the area where the mirror field stands was to about 5% north-south slope and 2% east-west whilst the buildings now stand on a flat area 214 m above the sea level.

After ENEL completed the acquisition of the site (15/10/78) the "Site preparation works" started with preparation of:
. access roads to the site
. provisional works and facilities area
. provisional works barracks (offices, canteen, etc.)
. provisional works connection to electricity and water networks.

An automatic meteorological station was installed at the site and started operation in January 1978.

Figure 3.8.1 - Location of EURELIOS

3.8.3 Description of works

The main civil works performed, outlined in Fig. 3.8.2, are the following:

- Mirror field. Ground preparation, mirror foundations, roads, draining system and fencing;
- Tower. Tower foundation, with an area of 100 m^2;
- Main building. The main building is composed of two rectangular blocks connected by a short corridor. Both buildings have steel frames, external tile curtain walls and metal deck roofing appropriately insulated and water-proofed. Offices, meeting room and a canteen are in the smaller, 250 m^2 block. After the corridor, the second block contains the machine room, battery room, air conditioning and boiler rooms, control and equipment rooms. This area is about 753 m^2 and height is between 4.70 and 9.50 m. In the machine room a travelling bridge crane is available to ease the handling of the heavier items of equipment. Adjoining the machine room on the south side, three external cells contain circuit breakers and two small transformers;
- Warehouse and workshop building. This building also has a rectangular plan and is constructed with steel frames, external tile curtain walls and roofing similar to that of the main building. It contains the electrical and mechanical workshops and the warehouse and is also served by a travelling bridge crane. The total floor area for warehouse and workshops is about 480 m^2;
- Cooling water basin and pumps cabin. Near the buildings is a group of water-related tanks, etc., including an open reinforced concrete tank with a base area of about 30 m^2 and volume about 80 m^3 (for used cooling water), tanks to separate oil contamination from waste water, a fibre glass tank for demineralised water for the steam cycle and a small building containing water discharge pumps;
- Finishing. Roads, car parking and other open storage areas are asphalted. The heliostat field, between access roads, is finished with large chippings and the remaining areas have been appropriately landscaped and planted. The entire drainage system (buildings, roads, etc. mirror field and other sources of waste water, after treatment) feeds into a common drain for discharge to the Simeto river, nearby. A photograph showing a general view of the plant during testing is reproduced in Fig. 3.8.3.

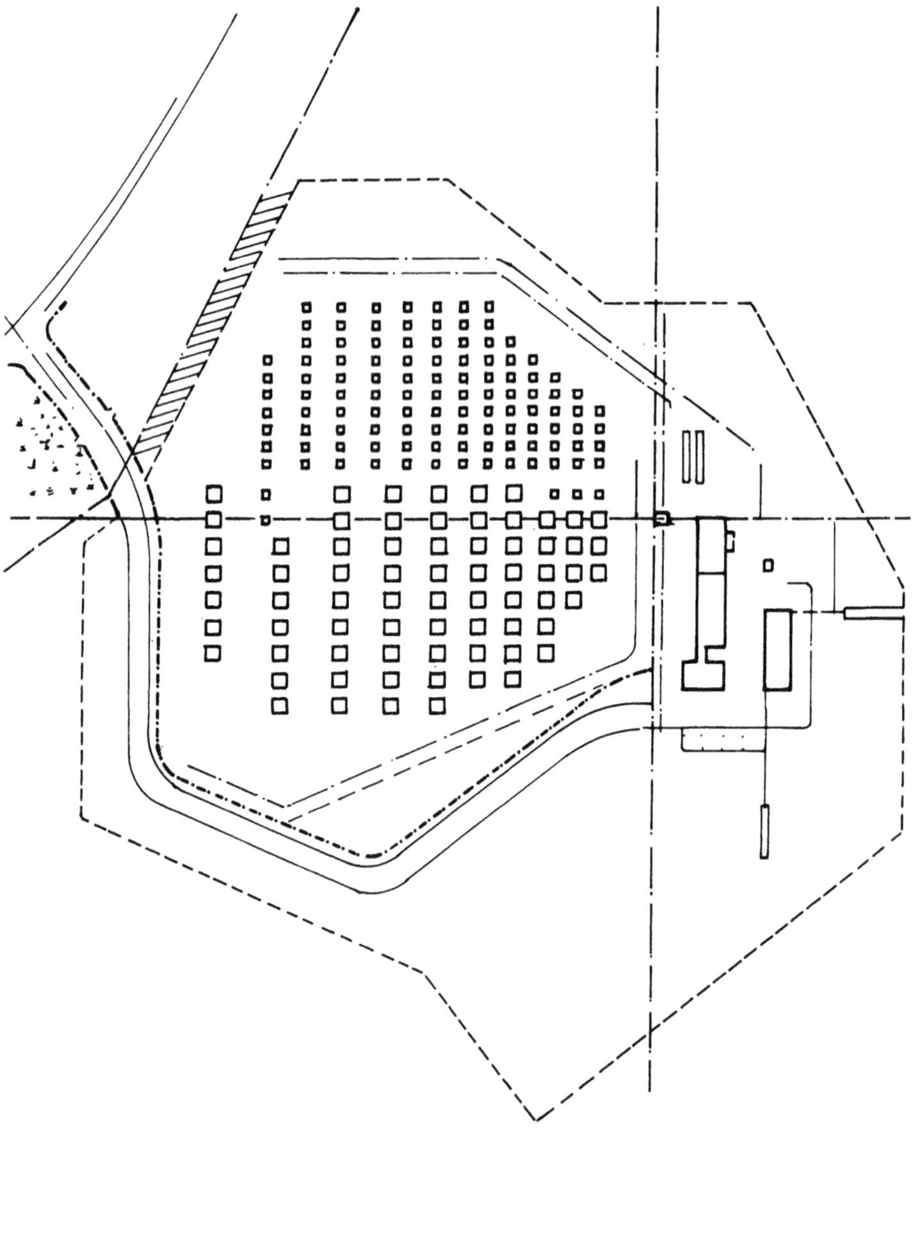

Figure 3.8.2 - General Lay-out

Figure 3.8.3 - General View during Testing

4. Operation of the Plant

4.1 INTRODUCTION

Detailed operation manuals have been produced by ENEL and ANSALDO and one important result of the future use of the plant will be the identification of amendments to the operation procedures, amendments which could improve the efficiency of such plants both thermoelectrically and in terms of manpower utilisation. This chapter, however, attempts to summarise the general operational principles and options as they appear in the Operation Manual as it now is.

4.2 OPERATING PROCEDURES

4.2.1 Morning start-up

Set-up the emergency feedwater system in start-up position and verify the correct positions of all relevant valves.
After verifying that the system is full of water and that the circulating pumps are in working order, start-up the cycle, pressurising the receiver and allowing water to flow through the receiver and the turbine bypass.
After flow is established and feedwater and receiver regulators and the emergency feedwater system have been put in automatic, the heliostats can be moved into the receiver (group by group at one minute intervals).
After about a half-hour of illumination, steam is produced and sent to the flash tank, the normal pressure of which is about 5 bars.
After a further half-hour, as the temperature at the receiver outlet passes 340°C, the turbine main header inlet can start being preheated. This takes about 20 minutes, after which the turbine is started and its speed increased under operator control at a standard rate of 200 rpm per minute until the synchronous speed of 8195 rpm is reached, at which point the alternator can be coupled to the grid. The total time from the pointing of the mirrors until the parallel is typically about two hours.
Once in parallel and with a receiver outlet temperature of about 470°C the plant can be put entirely under automatic control using a pressure set point appropriate to the time of year. The generator load is thereby adjusted to follow the fluctuations of solar flux.

4.2.2 Cold start-up with full radiation

The routine is virtually the same as for a morning start-up until the point at which the heliostats would have been moved from "standby" to "receiver". The required feedwater rate is first determined and set as a function of the observed and forecast pyroheliometer readings in the field and the helio-

stats are then moved one group every five minutes into the "receiver" position. The demanded feedwater flow rate is then adjusted according to the actual receiver fluxmeter measurements.
The procedure then reverts to that for morning start-up.

4.2.3 Hot start-up

This differs in concept from that of cold start-up in that, although the receiver is not initially illuminated, it is still pressurised with steam at at least 20 bar. The turbine may be shut-down or running on steam from the salt-storage system, the steam cycle pressure regulators generally having reverted to manual control.
Based on the observed insolation, appropriate setpoints are established and the bypass valve partially opened.
The mirror sequence "standby-to receiver" is initiated and the bypass valve put to automatic.
The receiver pressure is allowed to increase at about 1 bar/minute.
When the steam temperature is high enough, either

a) as soon as the steam pressure at the turbine inlet reaches 18 kg/cm^2 the automatic closure of salt storage steam valve is executed,
b) if the turbine is to be restarted, the hot start procedure is followed.

When the steam temperature reaches 470°C the whole system can be left in automatic mode.

4.2.4 Normal operation

After start-up the plant is automatically operated and the produced power is related to the daily insolation diagram. However, to maximise efficiency the receiver pressure set point may be adjusted periodically to keep the turbine inlet pressure valve about 70% open. Also it is possible, if necessary, to recharge the steam accumulator during the initial stage of normal operation.

4.2.5 Operation with cloud passing

When solar radiation drops below about 20% of nominal value for more than three minutes, the heliostat deviation sequence starts (receiver-to-standby) and if charged, the salt storage system is automatically put into operation. If the lack of sunlight is short enough for the heliostats not to reach standby, the plant will continue operating without operator involvement other than to send back to the "receiver" position any heliostat group which may have reached standby. The salt storage system will then not provide steam.
Steam starts to flow from the storage system as soon as the pressure to the turbine drops below the set point of the regulator, which should therefore be set as low as possible, commensurate with continued connection to the grid (typically 8 bar). Connection can then be maintained as long as sufficient steam is available from either source.
If all heliostat groups reach standby the operator prepares the plant, in manual mode, for the "hot start-up" procedure (Section 4.2.3).

4.2.6 Evening shut-down

Inhibit intervention of the salt storage system.
When the insolation drops below about 20% of nominal, the heliostat's

"receiver-to-standby" sequence is started automatically as above.
Put pressure regulation and receiver attemperator controls to manual and decrease the turbine load (by decreasing speed demand) as much as possible without breaking the coupling to the grid.
Decouple and reduce speed to a minimum.
Isolate the receiver and switch to"start-up" position to inhibit operation of the Emergency Feedwater System.
Stop the feedwater and booster pumps and put the entire system to "manual".

4.2.7 Emergency operations

Immediate emergency reaction to loss of pressure from the feedwater pumps is automatic and causes intervention of the emergency feedwater system. If this loss is due to failure of the auxiliary electrical supply only battery-operated functions will be available (CETHEL heliostat deviation and those items powered from ANSALDO's batteries, section 3.6.3) plus those which can be carried out with the residual service air. Otherwise normal services will continue to be available.
The mirrors are commanded out of the receiver, the receiver bottling valve is closed and its pressure regulated by the relief valve, N11-F300.
The regulation system continues to operate while service air is available, particularly to operate the feedwater regulating valves.
The operator:

- verifies that the emergency feedwater system operates correctly;
- verifies that the measured thermal input, Qer, to the receiver reduces correctly;
- when Qer is zero the operator closes the valves to avoid excessive thermal shock to the receiver.

The plant can then be normally shut-down.

5. Phase D

The contract for Phases B and C envisaged subsequent optimisation and improvement. A new two-year phase, Phase D, officially commenced on 5 January 1981.
Phase D is in two parts. The objective of Part I is to bring the plant up to full power and to establish a detailed work programme, DWP, for Part II.

A proposed DWP (dated June 14th, 1981) has, at the time of writing, been submitted by the Consortium. The main items of interest in the research programme planned for Part II, following the scheduled achievement of maximum power by the end of August, are:

1) plant dynamic behaviour and control system;
2) instrumentation and automation;
3) flux measurements in the receiver;
4) study of the mirror field behaviour;
5) analysis of the plant performance and optimisation of plant operation;
6) systems evaluation and final report.

These tasks are to be carried out by ENEL, CETHEL and MBB with assistance from Ansaldo should operational difficulties or other problems appear in those parts of the plant which Ansaldo supplied. Ansaldo, of course, still has its responsibility under the Phase C contract to install the Francia pyrex anti-radiation structure, deliberately delayed to allow quantitative experimental experience to be gained of the effects of such a structure on the performance of the receiver.

In addition to the work under the Consortium's Phase D contract, agreement has been reached between the Consortium, the Commission and Belgium for the latter to supply and install, by mid-1982, a Data Acquisition and Supervision System (DASS). A contract is being finalised between the Commission and Tractionel/Belgonucléaire. DASS will provide automatic and more comprehensive records of the operating plant than are at present available. GTS again has a direct contract from the Commission to provide advisory services in Phase D, particularly with respect to DASS, as it did in Phases B and C.

As for the future, provision was made in the layout of the heliostat field for slight enlargement. The receiver, turbine, generator, etc., were designed to accept the corresponding increase of power.

The possibility of some measure of integration between EURELIOS and proposed photovoltaic plant in an adjacent area has also been mooted - if only by co-ordination through an extended DASS.

Although other thermal cycles are not planned for the foreseeable future, the tower is compatible with the installation of different thermal cycles, making use of the existing heliostat field and other infrastructure of the plant. This type of development would, of course, require completely new negotiations with ENEL which would by that time presumably be full owners of the plant and using it as a routine, albeit small, contributor to the Sicilian electrical grid.

9 789027 717283

GPSR Compliance

The European Union's (EU) General Product Safety Regulation (GPSR) is a set of rules that requires consumer products to be safe and our obligations to ensure this.

If you have any concerns about our products, you can contact us on ProductSafety@springernature.com

In case Publisher is established outside the EU, the EU authorized representative is:

Springer Nature Customer Service Center GmbH
Europaplatz 3
69115 Heidelberg, Germany

Batch number: 10584448

Printed by Printforce, the Netherlands